Neopragmatic Horizontal Geographies

Volume 2

Series Editors

Central aspects of philosophical neopragmatism have been made available for integrative spatial research in recent years. Following classical pragmatism, which measures the quality of cognition by its suitability for problem solving, rather than by philosophical or moral principles, neopragmatism also integrates considerations of language theory. The central element of this meta-theoretical approach is the plurality of perspectives appropriate to the subject matter. This plurality gains particular relevance in the study of complex objects. Especially in the case of spatial syntheses, as they are carried out in 'horizontal geographies' (e.g. in the form of region, city or landscape), a multitheoretical, also multimethodical and inter- and transdisciplinary approach is required – following the neopragmatic approach. Due to the inter- and transdisciplinarity of the approach, there is a special focus on innovative forms of presenting the results of research, following the neopragmatic approach of horizontal geographies.

The aim of the series is first of all to elaborate the meta-theoretical fundamentals of 'neopragmatic horizontal geographies', which form the justifying basis for the triangulation of theories, research methods derived from them, and last but not least methods of representation. Studies that operationalize the meta-theoretical framework with different spatial as well as theoretical and methodological shear points illustrate the potentials of the approach for theory-driven horizontal geographic research.

Olaf Kühne · Karsten Berr · Dennis Edler ·
Petra Lohmann · Kai Schuster

Neopragmatism - Inverse Landscape - (Carto)graphic Representation

From Critique to Further Development

Springer

Olaf Kühne
Forschungsbereich Geographie
Eberhard Karls Universität Tübingen
Tübingen, Baden-Württemberg, Germany

Karsten Berr
Forschungsbereich Geographie
Eberhard Karls Universität Tübingen
Tübingen, Baden-Württemberg, Germany

Dennis Edler
Institut für Geographie
Ruhr-Universität Bochum
Bochum, Nordrhein-Westfalen, Germany

Petra Lohmann
Department Architektur, Fakultät II
University of Siegen
Siegen, Nordrhein-Westfalen, Germany

Kai Schuster
Fachbereich Gesellschaftswissenschaften
Hochschule Darmstadt
Darmstadt, Hessen, Germany

ISSN 2731-9881 ISSN 2731-989X (electronic)
Neopragmatic Horizontal Geographies
ISBN 978-3-031-89404-6 ISBN 978-3-031-89402-2 (eBook)
https://doi.org/10.1007/978-3-031-89402-2

This Springer imprint is published by the registered company Springer Nature Switzerland AG
The registered company address is: Gewerbestrasse 11, 6330 Cham, Switzerland

Competing Interests The authors have no competing interests to declare that are relevant to the content of this manuscript.

Contents

List of Figures

Chapter 1
Introduction

Inverse landscapes can be understood as a manifestation of landscape contingency. Landscape contingency arises from the fact that landscape is not only determined in terms of its social constructions and its individual allocations, but also in terms of its material foundations. Contingency is understood as that which is not requisite, yet also not impossible. Contingent landscapes are therefore landscapes that could be different in terms of social constructions, individual allocations, as well as in terms of their material foundations, but not arbitrarily different. The concept of inverse landscapes was originally developed in the early 2010s to illustrate the power-bound nature of landscape, namely to present what could not manifest itself, even if these manifestations had been possible, precisely for reasons of power (Kühne 2012, 2013a, 2013b). The concept was updated through a neopragmatist framing and the associated constructive turn (Kühne 2024b; Kühne, Berr, Koegst 2023a). The focus was not on what could not manifest itself as a result of power processes, but on what was innovative and empowering (Nussbaum 1999, 2009; Sen 1999); in short, in the sense of Ralf Dahrendorf (1979), the potential of creating and dealing with inverse landscapes.

This book joins the ranks of numerous works by authors who have dealt with the operationalization of neopragmatism (here especially in the tradition of Brandom 2001, 2023; Putnam 1990a, 1990b, 1997; Rorty 1980, 1997 [1989], 1999), which—to put it briefly—can be outlined as a synthesis of continental European philosophy of language and American (USA) pragmatism, for the spatial and, in particular, landscape sciences (for example: Berr und Kühne 2023; Chilla et al. 2015; Edler et al. 2023; Hinz et al. 2024; Kühne 2018; Kühne et al. 2021; Kühne 2024a; Kühne et al. 2024a; Kühne 2024b; Kühne und Jenal 2020; Kühne und Koegst 2023; Papendorf et al. 2024; Stemmer et al. 2023). In doing so, we wanted to make a contribution to the social scientific resonance that neopragmatism has experienced (e.g.,Olson 1989; Schulenberg 2021; Schulenberg 2021): Olson 1989; Schulenberg 2021; Selk und Held 2021; Topper 1995; Zoll 2023) within the spatial sciences (not only those that see themselves connected to the social sciences, such as natural science approaches).

O. Kühne et al., *Neopragmatism - Inverse Landscape - (Carto)graphic Representation*, Neopragmatic Horizontal Geographies 2, https://doi.org/10.1007/978-3-031-89402-2_1

Particularly when dealing with complex phenomena, neopragmatism enables a substantiated triangulation of different theories, methods, sources, and researcher perspectives, including the involvement of people without distinct technical expertise on the topic and—this is a central theme of this book—(carto)graphic representation. Compared to other theories (and metatheories), neopragmatism, as a result of its focus on the usefulness of results, is not geared towards replacing old paradigms with new paradigms, but rather towards weaving compartments of 'vocabularies' whose potential to interpret the world has diminished into redescriptions; in short, reconciling older "innovations with newer ones" (Rorty 2023, S. 252). Even if neopragmatism rejects representationalism and can be categorized as a more-than-representational theory, it is possible to use results from studies that have been obtained from a positivist perspective, for example. Even if neopragmatism ties in with a social constructivist tradition (Rorty 2023), neopragmatism nevertheless emphasizes pan-relationalist integrations. Self, language, and society emerge solely through their contexts, not through any 'being', as essentialism assumed. This also applies to landscapes, which are created by inner and outer contexts that are always reversible and can only be temporarily stabilized. This also gives rise to inverse landscapes, i.e., landscapes that are not to be found in this form (as social constructions, individual approaches, or material embodiments), but can be found in the space of contingencies, plus can also become landscapes from inverse landscapes if they find widespread social acceptance (an example: the old industrial landscape, which is generally accepted today, but was usually considered a non-landscape decades ago; see on this: Edler 2020; Höfer und Vicenzotti 2013; Kühne 2016; Kühne, Jenal, Berger 2023c; Schönwald 2015; Weber 2020).

This book is the result of a detailed discussion between the originator of the concept of inverse landscapes (Olaf Kühne) and the person who pursued the neopragmatist orientation of the concept together with the originator (Karsten Berr) alongside 'friendly critics', who, after various publications (see above) and oral presentations of the concept, asked questions from different perspectives and criticized it with the aim of further developing the concept in terms of theoretical foundation, differentiation, and presentation of results. These questions and criticisms came from the fields of philosophy (Petra Lohmann), psychology (Kai Schuster), and cartography (Dennis Edler). The initial idea of designing questions and criticism in a personalized scheme of question/criticism and answer was soon abandoned, as the 'critics' were soon concerned with the further development of the concept and the originators were committed to criticizing their own approach, with the result that the former critics defended the concept against the criticism of its originators and solutions to the problems raised were sought together. The structuring basis of the critical reference were categories of criticism that some of us had developed elsewhere (Kühne, Berr, Lohmann 2023b) and which at this point—as a side effect—are subjected to a review concerning their usefulness for further conceptual development. The criticism thus carried out ranges from internal criticism (as involves the examination of internal consistency), criticism of contextualization (whether the theory proves itself in the context of other results), metatheoretical criticism (which establishes a relationship to other theories), criticism from the perspective of other worldviews, and criticism that refers to the examination of consequences (including unintended side effects), to practical criticism, which—particularly relevant for a pragmatist approach—questions theories and methods with regard to their effects on the practice of life. In the critique, as well as the resulting further developments of neopragmatist landscape theory, the concept of contingent landscapes, inverse landscapes, and their (carto)graphic representations, a hermeneutic approach is followed in that we have detailed and expanded the critique, for example against the background of reactions to it, which has led to a corresponding further development of our conceptual considerations.

In the recent past, six levels of triangulation of neopragmatist research have been identified, whereby triangulation—following Davidson (2004)—is understood as the relationship of two elements (such as theories) against the background of a common framework (such as a metatheory) (Kühne et al. 2024b; Kühne 2024b; Kühne, Berr, Koegst 2023a; Kühne und Koegst 2023). The six levels include: *theory triangulation, method triangulation, researcher triangulation, triangulation of sources*, the *perspectives of people with different degrees of professional access to the topic*, and *triangulation of (carto)graphic representations.* The work presented here is conceptual. In other words, we are not conducting new empirical research, but rather subjecting the current (also empirically obtained) state of research to a critical examination. This in turn can serve as a basis for new empirical studies. In addition, we are broadening the professional horizon of neopragmatist landscape research by adding Kai Schuster, a psychologist and architect, to the team. As a result of the conceptual approach, in seeking greater clarity regarding aspects of neopragmatist landscape theory as well as the approach of contingent and inverse landscapes including their (carto)graphic representation (which have not been sufficiently reflected upon to date), we have not integrated people without specialist access to the topic into our research heuristics in this case. The level of representation, however, is of central importance, as we are clearly concerned with the consequences of the existential critique of positivism through critical cartography. In doing so, we further develop (carto)graphic methods of representation, on the one hand to be able to deal with contingency, and on the other to develop useful approaches to the world. Of the above-mentioned levels of triangulation, in this book we use those of theories, researcher perspectives, and representations. On this basis, we combine three objectives:

1. Reflection on the concept of inverse landscapes and their representations, which were developed from the neopragmatist principles of increasing contingency, especially with the stylistic device of irony.
2. Reflection on the conditions of possibility for the expansion of contingency, especially through irony, against the background of the unequal distribution of symbolic capital (Bourdieu 2005, 1983, 2016) and of life chances (Dahrendorf 1979).
3. Further development of the concept of inverse landscapes and their representations resulting from the reflection on the conditions of possibility for the expansion of contingency, especially through irony.

More abstractly formulated, the aim is to further develop the concept of inverse landscapes and, as a result of critical considerations, to expand its potential to provide solutions to problems and possibilities for expanding contingency, while at the same time pointing out conceptual challenges and finding ways of dealing with them.

After this introduction to the topic and the aims of the work, we turn our attention in Chapter 2 to the presentation of the quite extensive theoretical foundations of our book. This ranges from the notion of neopragmatism as a metatheoretical foundation, its operationalization in irony as an expression of the awareness of contingency, the notion of our quasi-ontological approach to landscape (based on the theory of three landscapes derived from Karl Popper's 'three worlds theory'), to the integration of philosophical, sociological, and psychological theories that have proven useful—or that we hope will prove useful—in understanding the complex subject of 'landscape'.

Subsequently, in Chapter 3, we discuss the concept of inverse landscapes as a concretization of landscape contingency. The (carto)graphic representation of inverse landscapes is also addressed. In the following Chapter 4, we turn to the previously mentioned framework of critique. The different types of critique are presented here, which in turn are discussed in Chapter 5 via the concept of inverse landscapes and their (carto)graphic representations, as well as their theoretical foundations, particularly the neopragmatist landscape theory and the contingent landscape approach. Here, the six types of critique are each applied to six thematic aspects, (a) preconditions for a neopragmatist landscape theory, (b) neopragmatist understandings of landscape themselves, (c) the concept of inverse landscapes, (d) the possibilities of inverse landscapes to operationalize contingency, (e) contemplating inverse landscapes and their (carto)graphic representations ironically, and (f) integrating inverse landscapes into the context of an understanding of the world, in general, and space, in particular, that is shaped by emotion and compassion. In Chapter 6, these six thematic aspects are taken up on the basis of the six categories of criticism in each case and the criticisms put forward are addressed. These are taken as an opportunity to detail or expand the theoretical or conceptual considerations and to integrate or reject the criticism. In Chapter 7, the results are summarized and classified in the context of neopragmatist landscape research and open questions are identified.

References

Berr, K. & Kühne, O. (2023). Aesthetic constructions of landscape between society, individual and objects. A neopragmatic approach. *Studi di estetica LI, IV Serie* (26), 1–29. https://doi.org/10.7413/1825864638.

Bourdieu, P. (2005 [1983]. Ökonomisches Kapital – Kulturelles Kapital – Soziales Kapital. In P. Bourdieu (Hrsg.), *Die verborgenen Mechanismen der Macht* (S. 49–80). Hamburg: VSA.

Bourdieu, P. (2016). *La distinction: Critique sociale du jugement* (Le Sens commun). Paris: Editions de Minuit.

Brandom, R. B. (2001). *Making it explicit. Reasoning, representing, and discursive commitment* (4. print). Cambridge, Mass: Harvard Univ. Press.

Brandom, R. B. (2023). Vollendung der Aufklärung. Vorwort von Robert B. Brandom. In E. Mendieta (Hrsg.), *Pragmatismus als Antiautoritarismus* (S. 7–41). Berlin: Suhrkamp.

Chilla, T., Kühne, O., Weber, F. & Weber, F. (2015). „Neopragmatische" Argumente zur Vereinbarkeit von konzeptioneller Diskussion und Praxis der Regionalentwicklung. In O. Kühne & F. Weber (Hrsg.), *Bausteine der Regionalentwicklung* (S. 13–24). Wiesbaden: Springer VS.

Dahrendorf, R. (1979). *Lebenschancen. Anläufe zur sozialen und politischen Theorie* (Suhrkamp-Taschenbuch, Bd. 559). Frankfurt (Main): Suhrkamp.

Davidson, D. (2004). *Subjektiv, intersubjektiv, objektiv*. Frankfurt (Main): Suhrkamp.

Edler, D. (2020). Altindustrielle Klanglandschaften und moderne 3D-kartographische Ansätze ihrer Wiedergabe und Erhaltung. In R. Duttmann, O. Kühne & F. Weber (Hrsg.), *Landschaft als Prozess* (S. 267–280). Wiesbaden: Springer VS.

Edler, D., Kühne, O., Keil, J., Korte, A., Weißmann, M., Galla, C., Zilt, L. & Dickmann, F. (2023). Geodatenkultur der Diversität: eine neopragmatische Skizzierung. *KN – Journal of Cartography and Geographic Information 73*, 317–333. https://doi.org/10.1007/s42489-023-00146-6

Hinz, L., Weber, A.-M., Koegst, L. & Kühne, O. (2024). A Neopragmatic Perspective on the Processual Nature of Landscape – Coastal Land Loss in Louisiana in the Context of Scientific Findings, Social Patterns of Interpretation, and Individual Experience. *Sustainability 16* (5), 1–26. https://doi.org/10.3390/su16052078

Höfer, W. & Vicenzotti, V. (2013). Post-industrial landscapes: evolving concepts. In P. Howard, I. Thompson & E. Waterton (Hrsg.), *The Routledge Companion to Landscape Studies* (S. 405–416). London: Routledge.

Kühne, O. (2012). *Stadt – Landschaft – Hybridität. Ästhetische Bezüge im postmodernen Los Angeles mit seinen modernen Persistenzen*. Wiesbaden: Springer VS.

Kühne, O. (2013a). Landschaft zwischen Objekthaftigkeit und Konstruktion – Überlegungen zur inversen Landschaft. In D. Bruns & O. Kühne (Hrsg.), *Landschaften: Theorie, Praxis und internationale Bezüge. Impulse zum Landschaftsbegriff mit seinen ästhetischen, ökonomischen, sozialen und philosophischen Bezügen mit dem Ziel, die Verbindung von Theorie und Planungspraxis zu stärken* (S. 181–193). Schwerin: Oceano Verlag.

Kühne, O. (2013b). Landschaftsästhetik und regenerative Energien – Grundüberlegungen zu De- und Re-Sensualisierungen und inversen Landschaften. In L. Gailing & M. Leibenath (Hrsg.), *Neue Energielandschaften – Neue Perspektiven der Landschaftsforschung* (S. 101–120). Wiesbaden: Springer VS.

Kühne, O. (2016). Transformation, Hybridisierung, Streben nach Eindeutigkeit und Urbanizing former Suburbs (URFSURBS): Entwicklungen postmoderner Stadtlandhybride in Südkalifornien und in Altindustrieräumen Mitteleuropas – Beobachtungen aus der Perspektive sozialkonstruktivistischer Landschaftsforschung. In S. Hofmeister & O. Kühne (Hrsg.), *StadtLandschaften. Die neue Hybridität von Stadt und Land* (S. 13–36). Wiesbaden: Springer VS.

Kühne, O. (2018). Reboot "Regionale Geographie" – Ansätze einer neopragmatischen Rekonfiguration "horizontaler Geographien". *Berichte. Geographie und Landeskunde 92* (2), 101–121.

Kühne, O. (2024a). Neopragmatische Landschaftsforschung. In O. Kühne, F. Weber, K. Berr & C. Jenal (Hrsg.), *Handbuch Landschaft* (2. Aufl., 261–273). Wiesbaden: Springer VS.

Kühne, O. (2024b). *Redescribing Horizontal Geographies. A Neopragmatist Approach to Spatial Contingency, Complexity, and Relationships*. Cham: Springer International.

Kühne, O. & Jenal, C. (2020). *The Multivillage-Metropolis Baton Rouge. A Neopragmatic Landscape Biographical Approach on Spatial Pastiches, Hybridization, and Differentiation*. Wiesbaden: Springer VS.

Kühne, O. & Koegst, L. (2023). *Land Loss in Louisiana. A Neopragmatic Redescription*. Wiesbaden: Springer VS.

Kühne, O., Koegst, L., Zimmer, M.-L. & Schäffauer, G. (2021). "… Inconceivable, Unrealistic and Inhumane". Internet Communication on the Flood Disaster in West Germany of July 2021 between Conspiracy Theories and Moralization – A Neopragmatic Explorative Study. *Sustainability 13* (20), 1–23. https://doi.org/10.3390/su132011427

Kühne, O., Berr, K. & Koegst, L. (2023a). Contingency and Landscape. Basic Considerations on Graphic and Cartographic Representations in Recourse to the Concept of Inverse Landscapes as a Contribution to Deviant Cartographies with Examples on Louisiana. *KN – Journal of Cartography and Geographic Information,* 1–12. https://doi.org/10.1007/s42489-023-00145-7

Kühne, O., Berr, K. & Lohmann, P. (2023). *Landschaft zwischen Philosophie und Sozialwissenschaften. Eine Kritik.* Wiesbaden: Springer Fachmedien; Springer VS.

Kühne, O., Jenal, C. & Berger, S. (2023). Saarbrücken-Brebach – phänomenologische Zugänge zu einem touristisch kaum präsenten, altindustriell geprägten Ort. In O. Kühne, T. Freytag, T. Sedelmeier & C. Jenal (Hrsg.), *Landschaft und Tourismus* (RaumFragen, 613-639). Wiesbaden: Springer.

Kühne, O., Koegst, L. & Berr, K. (2024a). *Oilscapes of Louisiana – Neopragmatic Reflections on the Ambivalent Aesthetics of Landscape Constructions.* Wiesbaden: Springer Fachmedien.

Kühne, O., Koegst, L. & Berr, K. (2024b). *Oilscapes of Louisiana. Neopragmatic Reflections on the Ambivalent Aesthetics of Landscape Constructions.* Wiesbaden: Springer VS.

Nussbaum, M. C. (1999). *Gerechtigkeit oder Das gute Leben.* Frankfurt (Main): Suhrkamp.

Nussbaum, M. C. (2009). *Frontiers of justice. Disability, nationality, species membership.* Cambridge, London: Harvard University Press.

Olson, G. A. (1989). Social Construction and Composition Theory: A Conversation with Richard Rorty. *Journal of Advanced Composition 9* (1/2), 1–9.

Papendorf, K., Schumacher, T. & Güntherodt, J. (2024). The Scramble for Europe. A Neopragmatic Approach to a Deviant Scenario for Colonization. In F. Papadimitriou & O. Kühne (Hrsg.), *Deviant Landscapes. A Journey to Exotic and Futuristic Landscapes and Spaces* (in this anthology). Wiesbaden: Springer VS.

Putnam, H. (1990a). *Die Bedeutung von „Bedeutung"* (2., durchgesehene Auflage). Frankfurt (Main): Klostermann.

Putnam, H. (1990b). *Vernunft, Wahrheit und Geschichte.* Frankfurt (Main): Suhrkamp.

Putnam, H. (1997). *Für eine Erneuerung der Philosophie* (Universal-Bibliothek, Bd. 9660). Stuttgart: Reclam.

Rorty, R. (1980). *Philosophy and the mirror of nature* (1. paperback print). Princeton, N. J.: Princeton University Press.

Rorty, R. (1997 [1989]. *Contingency, Irony, and Solidarity.* Cambridge: Cambridge University Press.

Rorty, R. (1999). *Philosophy and social hope* (Penguin books, 1. publ). London: Penguin Books.

Rorty, R. (2023). *Pragmatismus als Antiautoritarismus.* Berlin: Suhrkamp.

Schönwald, A. (2015). Die Transformation von Altindustrielandschaften zwischen Kontinuität und Wandel. In O. Kühne, K. Gawroński & J. Hernik (Hrsg.), *Transformation und Landschaft. Die Folgen sozialer Wandlungsprozesse auf Landschaft* (S. 63–73). Wiesbaden: Springer VS.

Schulenberg, U. (2021). Zwischen Liberalismus und Marxismus. Rortys Pragmatismus im politischen Kontext. In V. Selk, C. Held & T. Schwuchow (Hrsg.), *Soziale Hoffnung, liberale Ironie. Zur Aktualität von Richard Rortys politischem Denken* (Staatsverständnisse, Bd. 145, S. 25–50). Baden-Baden: Nomos.

Selk, V. & Held, C. (2021). Schlechte Aussichten. Richard Rortys politisches Denken in der Krise des demokratischen Liberalismus. In V. Selk, C. Held & T. Schwuchow (Hrsg.), *Soziale Hoffnung, liberale Ironie. Zur Aktualität von Richard Rortys politischem Denken* (Staatsverständnisse, Bd. 145, S. 159– 178). Baden-Baden: Nomos.

Sen, A. (1999). *Development as Freedom.* Oxford: Oxford University Press.

Stemmer, B., Bernstein, F., Behre, E. & Kaußen, L. (2023). Naherholung als Teil der grünen Infrastruktur – ein neopragmatischer Ansatz. In O. Kühne, T. Freytag, T. Sedelmeier & C. Jenal (Hrsg.), *Landschaft und Tourismus* (RaumFragen, 253-275). Wiesbaden: Springer.

Topper, K. (1995). Richard Rorty, Liberalism and the Politics of Redescription. *American Political Science Review 89* (4), 954–965. https://doi.org/10.2307/2082520

Weber, F. (2020). ‚Über Tage' – das Saarpolygon als Anker sich wandelnder Altindustrielandschaften im Saarland. In R. Duttmann, O. Kühne & F. Weber (Hrsg.), *Landschaft als Prozess* (S. 377–402). Wiesbaden: Springer VS.

Zoll, P. (2023). Rorty und „post-faktische" Politik. In M. Müller (Hrsg.), *Handbuch Richard Rorty* (1st ed. 2023, S. 997–1016). Wiesbaden: Springer Fachmedien.

Chapter 2
Propaedeutic Course on Inverse Landscapes: Basic Contingency Considerations

As has already become clear from the introduction, neopragmatism has not only been able to establish itself as a philosophical movement but has also found resonance in the social and spatial sciences. A key strength of neopragmatism lies in its versatility and capacity for integration. It can be seen as a metatheoretical framework for integrating different theories, additionally, as a pan- relationalist theory it offers connections to related theoretical approaches.

In this chapter, we will first deal with neopragmatism as a metatheoretical approach and then turn to the significance of irony in the neopragmatist perspective, a topic that will be pursued again and again throughout the rest of the work. We will then look at the three world theory, which we understand as quasi-ontological, and its applications for landscape research, which has proven itself as an analytical tool as regards the neopragmatist metatheoretical approach involving spaces and landscapes when it comes to understanding approaches (whether routine, methodological, or theorctical) and their relationships. We will then look at the relationship between 'dynamis' (capacity, power, possibility) and 'energeia' (activity, reality) against the background of the experience and production of contingency, notably that of landscape. Having laid these foundations from a primarily philosophical and scientific theory perspective, we will now turn to psychological and sociological approaches to contingency and landscape. These in turn are linked back to the neopragmatist concept of irony, which is particularly important in the presentation of results. Bringing together the multi-perspectivity of this chapter, we conclude with a summary.

O. Kühne et al., *Neopragmatism - Inverse Landscape - (Carto)graphic Representation*, Neopragmatic Horizontal Geographies 2, https://doi.org/10.1007/978-3-031-89402-2_2

2.1 The Neopragmatist Metatheoretical Approach

Neopragmatism can be categorized in the series of anti-essentialist and anti-metaphysical theoretical traditions, such as post-structuralism, existentialism, hermeneutics, postmodern holism, Wittgensteinianism, and others (Rorty 2018 [1994]). Briefly summarized, it forms an emergence (or in neopragmatist terminology: a redescription) of American pragmatism (James, Peirce, Mead, Dewey) and continental European philosophy of language (Wittgenstein, Derrida, Foucault).

The aim of pragmatism is not to find irrefutable and timeless (essentialist) truths (whose existence it doubts in the first place), but to produce useful knowledge for concrete problems (for a well-founded introduction, see: Bernstein 2010). The measure of usefulness is—Rorty (2023, p. 62) following Dewey and James—a utilitarian one, after all, "the only moral or epistemological criteria we have or need [...] ultimately boil down to the question of whether the performance of an action or the advocacy of a belief contributes to the increase of human happiness in the long run". The idea of the development of human society, which is committed to such a goal, cannot be teleological, but follows an evolutionary incrementalism, which, as with Rorty (2023, p. 65) from the efforts of James and Dewey to "harmonize philosophy with Darwin by placing human striving for truth and the good on the same level as the activities of the lower animals, that is, the development of culture on the same level as biological evolution". From this perspective, philosophical and scientific progress is "gained within the framework of practical intersubjectivity and not through individual and a priori acts of consciousness, because even logical conclusions are based [...] on a social principle" (Schubert et al. 2010, p. 23). Neopragmatism rejects the idea of analytical philosophy that "only natural science describes the world in its being-in-itself and independently of a particular perspective" (Putnam 1997, p. 8) by pointing to the inescapability of the non-scientific horizon of experience.

Neopragmatism extends the critique by addressing the question of the "connection between language and the world" (Putnam 1997, p. 8).

The aim of a neopragmatist approach to the world is to "replace the distinction between appearance and reality with the distinction between more or less useful descriptions of the world and ourselves" (Rorty 2018 [1994], p. 16). Neopragmatism, therefore, does not claim that there is no world outside of language or consciousness (it is therefore not a solipsistic philosophy), it only considers it impossible to depict the world outside of language (Rorty 1980). Following this reasoning, Rorty (2023, p. 54) rejects the idea that there can be only 'the one' correct description of a thing: "There are only those descriptions which, by relating it to other things, embed it in contexts that serve our present, multiple needs". This has fundamental implications for the relationships between humans and 'the world out there' because these are formed "not in a representational but in a causal way.

The world can be the cause of the fact that, as soon as we speak a language, we arrive at certain beliefs. These are developed in response to the causal stimuli of the world in order to be able to act successfully in it" (Heindl 2023, p. 10). Here, the constitutive reference of pragmatism to Darwin's theory of evolution is actualized when Rorty (2003, p. 215) states that "just as the brute force of the environment has led to various successive stages of biological evolution, it has also brought about the successive stages of cultural evolution". In neopragmatist terms, the pressure of the environment (especially social pressure) leads to an evolution of vocabularies involved. Rorty (1997 [1989]) understands this to mean common platforms of social justification practice. These vocabularies, in turn, are exclusively relational and are based on the fact that "words derive their meaning from their use in connection with other words" (Buschmeier 2023, p. 3). At this point it becomes clear again that a representational relationship with the world for the constitution of 'truth' is not possible outside of vocabularies: "Since truth is a property of sentences, since the existence of sentences is dependent on vocabularies, and since vocabularies are made by people, the same is true of truths" (Rorty 1989, p. 49). This in turn has the consequence that descriptions of the world are contingent, because there is not "only one 'correct' version of the world, but a whole number of 'correct versions'" (Putnam 1997, p. 143). However, such a recognition requires a metaperspective, because what is considered 'truth' within individual vocabularies, as the 'correct version of the description of the world', is dependent on the justifications accepted within the vocabulary (Rorty 1999, 2003). In justification, the speaker "draws on a network of convictions that is formed in the communicative process of agreement and contradiction, of justification and correction of convictions, and not through introspection that encounters an apparently pre-social foundation of knowledge in the subject" (Kronenberg 2014, pp. 46–47). The different vocabularies in turn form "mutually distrustful (and *not*: incomprehensible) communities of justification" (Rorty 2023, p. 178; emphasis in the original). This contouring of one's self within the otherness of the alternative vocabularies has an exclusionary effect: "The other 'does not belong to us', he does not belong to the *real* people, the *paradigmatic* people, i.e., to those who are to be treated with respect with regard to their person and their opinions" (Rorty 2023, pp. 178–179; emphasis in the original). This distinctive devaluation of alternative vocabularies, in turn, results in the formation of an identity within one's own vocabulary. The differentiation of disparate vocabularies, which was initially described as a "phenomenon of linguistic division of labor" (Putnam 1990a, p. 43) has led to a reciprocal discursive closure between the individual language communities.

At the level of these mutually exclusive vocabularies, mediation is therefore very difficult; in the following we will deal principally with scientific vocabularies, i.e., research programs (Lakatos 1974). A metalevel reflection, as we will provide below, offers a way out. The starting point is the realization that it is not enough "to recognize that a multitude of other vocabularies exist alongside one's own; rather, one must relate this observation to anti-fundamentalist conceptions of language, man and the world in order to be able to make the diagnosis of contingency" (Deines 2008, p. 158). Scientific vocabularies are therefore contingent. The need to justify one's own research program therefore arises from the diagnosis of its contingency (Rorty 1981) because no research program is self-evident or 'natural'; as a tool for solving tangible problems, it is always subject to the obligation to prove its usefulness. The insight into contingency for the sciences, according to Barnes (2023, p. 3), means creating new vocabularies: "By inventing new words, telling new stories and putting those words and stories into motion, it is sometimes possible to break through old stand-offs and logjams and repair sites where earlier conversation lagged or stalled or was abandoned". It is not a linear extension of traditionalism that constitutes science, but the creation and testing of new contingent interpretations in the form of redescriptions that "help to reveal the sheer contingency of our concepts and perspectives. Once a person recognizes the contingency of her inherited final vocabulary and other final vocabularies, she is more inclined to experiment with different descriptions or conceptions of herself and others and alternative ways of life" (Frazier 2006, pp. 462–463). The redescription of the world is always historically contextualized: "'Redescription' is an incessant attempt to express something new in the words of others and at the same time to recognize that what is said can only ever be parasitic" (Tautz 2023, p. 10).

Redescriptions thus arise from vocabularies that have become 'time-honored'. A vocabulary becomes 'time-honored' when its usefulness for interpreting the world has become fragile, for example, as an outcome of social developments. However, in this idea, 'time-honored vocabularies' are not simply replaced by new ones as is the case with the idea of 'paradigm shifts' (Kuhn 1970) but rather examines whether parts of 'time-honored' vocabularies still promise benefits. These promising portions are woven into new vocabularies: a redescription takes place. Not only does a 'time-honored' vocabulary have to be transferred into a new one, but several competing vocabularies can also be merged into one redescription (Calder 2007; Müller 2014; Rorty 1997 [1989])—a process that neopragmatism, in its combination of philosophy of language and pragmatism, has also accomplished. The thrust of this idea becomes clear in Rorty (following Dewey) when he states that, as mentioned earlier in the introduction, it is a matter of "reconciling older innovations with newer ones" (Rorty 2023, p. 252). Thus, every human achievement is "nothing more than a steppingstone to a greater achievement. We will never find descriptions that are so perfect that imaginative description loses its meaning" (Rorty 2019, p. 193). At this point, the central difference to post- structuralist philosophy of language becomes clear, which in turn arises from the utility orientation of the pragmatist roots of neopragmatism. Hence, for Rorty (1993, p. 96), resides the problem of post- structuralism: "When philosophy has finished showing that everything is social construct, it does not help us decide which social constructs to retain and which to replace". This is where the recourse to pragmatism comes in.

From the above, three criteria for the suitability of neopragmatist redescriptions, especially scientific, can be determined (Kühne et al. 2024). First, at the time of their creation (!), they must be more suitable for interpreting the 'world' than the 'traditional' vocabularies they incorporate. Second, they must be designed in such a way that they can be integrated into the fabric of future redescriptions (i.e., they must not claim to have found a final truth, such as essentialism). Third, they must interpret the world contingently in two regards. First, they must describe contingencies, but they must also produce contingent interpretations. From this metatheoretical position follows an instrumental relationship to the tools that science provides: theories, methods, data, and methods of representation. These can be triangulated with each other if it appears useful and appropriate for dealing with the research question—justified. Triangulation can be understood in neopragmatist terms as the relationship of at least two elements to each other within a broader framework (Davidson 2004 as an example: the mutual reference of two theories in a metatheoretical framework, such as neopragmatism). Especially when dealing with complex topics, the combination of different disciplinary and—when dealing with the everyday world—also everyday world perspectives seem to make sense. The everyday world has a threefold significance for neopragmatist science: All scientific knowledge is based on 'pre-understandings' of the world, which are proto-theoretical in nature.

Second, recourse to everyday language also makes it possible to establish communicability between different scientific vocabularies (Putnam 1990a). Recourse to everyday language also means evaluating the usefulness of speech descriptions outside one's own scientific vocabulary.

In terms of truth theory, the neopragmatism represented here can be assigned to coherence theories and consensus theories. Redescription refers to "the coherence and adequacy of a statement or a whole system of statements (a theory or a conceptual scheme) that makes them relationally acceptable, that is, the coherence of 'theoretical' or not too experience-bound beliefs with each other and with respect to more experience-bound beliefs, as well as the coherence of experience- bound beliefs with theoretical ones" (Putnam 1990b, p. 82). A prerequisite for the attestation of coherence, however, is a consensus on the specific construction of 'truth'. This consensus is established within a discourse community—in this case a scientific vocabulary (in the case of neopragmatism, a meta-vocabulary).

After this meta-stage reflection, which has as its central consequence the determination of the redescriptive capacity of research programs and theories formulated in vocabularies, we will deal with the operationalization of contingency and irony (also their connection) in the process of neopragmatist spatial research.

The central concept of philosophy by Richard Rorty (1997 [1989] is contingency, i.e., that which is neither necessary nor impossible. 'Contingency' stands "at one end of a set of oppositions whose other term is *variously necessary, essential, intrinsic*, or *unconditional*" (Topper 1995, p. 958; original formatting and spelling). History in general, and the development of philosophy in particular, is characterized by contingency: "Rorty's stress on the historical nature of philosophical problems—his 'historicism'—is designed to dispel the air of 'inevitability' that surrounds them. If, as it could have done, history had followed a different path, problems that now look as if they have forced themselves upon us by the sheer weight of their intrinsic intellectual difficulty and importance might never have surfaced" (Malachowski 2014, p. 17). Rorty finds the expression of the awareness of the contingency of language, the self, and society in his figure of the 'liberal ironist' (Reese-Schäfer 2016). This can be described as a person "who faces the fact that her central beliefs and needs are contingent" (Rorty 1989, p. 14). Liberal ironists can be described as people who have abandoned the idea that needs and hopes can be traced back to a deeper cause and instead hope "that suffering will diminish, that the humiliation of people by people will perhaps cease" (Rorty 1989, p. 14). The ironic attitude arises from an awareness of the contingency of language, self, and society, therefore meaning "never quite able to take themselves seriously because of a constant awareness that the terms in which they describe themselves are subject to change" (Rorty 1989, p. 128). Aware of the temporal and comprehensive limitations of vocabulary, the liberal ironist's "greatest fear [is] to get stuck in a vocabulary. Conversely, learning new vocabularies leads to a sensitization of perception, a broadening of horizons and a deepening of experience. In this sense, it is a game with language and existence" (Weber 2023, p. 7). Trust in irony stems from the insight that "there is nothing that we can rely on but ourselves and our fellow human beings" (Bernstein 2008, p. 22). The awareness of contingency not only means "mastering the contingency of one's own hermeneutic situatedness, but also recognizing it and taking responsibility for it" (Ramberg 2011, pp. 92–93). This is therefore about enabling other people to create themselves. The task of the liberal ironist thus extends beyond the realm of the private into that of the public. For Rorty, the realm of the private initially characterizes an area that "concerns questions of personal self-creation, [while] the realm of the public concerns our relationships to other people" (Kretz 2023, p. 5). Private self-creation becomes a necessity because the idea of an identitary 'core essence' of the human being does not exist, meaning that the 'self' must always be developed relationally. The private sphere of self-creation stands alongside the sphere of public solidarity: In the sense of neopragmatism, solidarity can be understood as an operationalization of the principle of 'fraternity', "it unites assistance with legitimate interests and justified claims" (Hacke 2021, p. 90). This humanity arises from a basic moral attitude of compassion (Rorty 1997 [1989], 1999). Rorty perceives a culture that is based on compassion and has become humane (2023, p. 55) as "conceiving of our moral task in terms of expanding our moral community and, through inclusion, ensuring that more and more diverse people fall under our use of the word 'we'". As the opposite of irony, however, is "common sense. It is the slogan of those who describe everything important unbiasedly in terms of the final vocabulary to which they and their environment are accustomed" (Rorty 1989, p. 128). Here, 'common sense' can be understood as a conclusive vocabulary in which existing patterns of action and convictions are understood as sufficient to approach, describe, and judge oneself, people, and the world. This approximation, description, and especially judgment also applies to those who strive to create alternative vocabularies (Rorty 1989). The background to Rorty's criticism of common sense, as understood in this way, is primarily a tradition that defines final vocabularies by referring to immovable truths and a "single permanent reality" that precedes language (Rorty 1989, p. 129) which underlies the many possible manifestations as essential. Rorty's second motive consists in the rejection of a chauvinism that often goes hand in hand with a claim to common sense. Another positive understanding of common sense, for example, in Aristotle, David Hume, George Edward Moore, and Charles Sanders, emphasizes the importance of its role as 'common sense', for example, as a critical corrective to exaggerated theoretical trains of thought, concepts, and theories. Georg Wilhelm Friedrich Hegel differentiated this situation of ambivalent evaluation to the effect that common sense is prone to contradiction and error in philosophy and the sciences and is therefore inadequate, whereas in practical everyday life it is often more appropriate than theoretical thinking when it comes to practical challenges.

An examination of 'landscape'—what we understand by this is the subject of Sect. 2.3—from a neopragmatist perspective deals with contingency in two ways: First, it clarifies the contingency of the world and, second, it adds contingent descriptions to it with its redescriptions. The awareness of contingency finds expression in these speech descriptions in an ironic attitude (Rorty 1997 [1989]). In this respect, our engagement with inverse landscapes and their (carto)graphic representations is connected to 'ironistic theorizing', i.e., it is based on a theoretical practice "that emphasizes its own contingency" (Rorty 2001, p. 27). But before we turn to the subject of landscape, let us turn our attention to irony, which has been and continues to be the subject of controversy, and which is central to neopragmatism.

2.2 Irony as an Epistemological Tool

The ironic approach to oneself, as well as to the world outside one's own consciousness, forms a core of Richard Rorty's neopragmatism (1997 [1989]), which guides us in our understanding of 'ironic cartography'. The beginnings of the word 'irony' can be traced back to around 400 BCE in ancient Greece. Irony means that the speaker expresses something other than that which he actually perceives, but it also means that the speaker deviates from the 'core of truth', not in a boastful way, towards 'too much', but in a modest way towards 'too little' (Aristoteles 2001 [ca. 335 – 322 BCE]. The effect of irony is not unambiguous; for example, it can be hurtful to people to whom irony is applied, it can be outrageous if it is directed at a serious issue. However, it can also, in the tradition of Socrates, advance knowledge processes in the form of modest questioning (Wirth 2017). A tradition that is continued in German Romanticism, whereby Schlegel (1963) emphasizes the presuppositional character of irony, because only those who have learned to deal with it will not be confused by it.

With Schlegel, irony can also be generalized from the concrete situation to a permanent role or even attitude (Schlegel 1963). This questions what is taken for granted (by others) and thus contributes to the 'deconstruction' of generally shared normality (Wirth 2017). This idea can also be found in Richard Rorty when he emphasizes the importance of irony for the awareness of contingency (Rorty 1997 [1989]), which we will look at in more detail later.

Terms such as 'irony', 'wit', 'comedy', 'humor', and other related expressions initially bring to mind something 'funny', 'amusing', or 'jocular', (in an exaggerated way) 'ridiculous', pejorative mockery, and derision. If the literature on these and other terms is examined, particularly in philosophy as well as literary and cultural theories, the term 'comedy' "now serves as a generic term for amusing things of various kinds, from the witty to the farcical and nonsensical to the satirical or humorous" (Kindt 2017b, p. 2; see also Grau 2010). 'Comic' and 'comical' can be traced back etymologically to the ancient comedy and denote objects of various kinds "to which an amusing effect is attributed" (Kindt 2017b, p. 2). In this context, *humor* can generally be defined as an attitude of people "that consists of an open-mindedness towards the comical" (Kindt 2017a, p. 7). The *joke* can be defined as a "concise, joking statement, usually in the form of a funny (short-) story that ends with a punch line" (Willer 2017, p. 11). *Irony* does not fit into this component of meaning of the comical and amusing without further ado, but rather refers to another strand of meaning that also concerns humor and wit and indicates an epistemological and epistemic function of critique. Gottfried Gabriel has reminded us of this function, which was particularly developed in the context of the rationalist epistemology of the Enlightenment by philosophers such as Christian Wolff, H. G. Baumgarten, and Georg Friedrich Meier (Gabriel 1996).

In this philosophical tradition, a distinction between 'aesthetic wit' and 'logical acumen' comes into play, which can also be informative and guiding for the following considerations on wit, humor, and irony. In this tradition, wit (lat. *ingenium*) stands for "*similarities in what is different*", while acumen (lat. *acumen*) stands for "*differences in what is similar*" (Gabriel 1996, p. 1; original formatting and spelling). Wit stands for "analogical thinking in family resemblances" in Wittgenstein's sense, which transcends existing conceptual boundaries, while acumen stands for "logical thinking in 'sharply defined' concepts" in Frege's sense, which can see and draw new conceptual boundaries. Gabriel refers to J. F. Fries, who has already directed this distinction to the difference between "the search for the *identity of the different* and the search for the *difference of the identical*" (Gabriel 1996, p. 3; there with reference to Fries). Parmenides (identity) and Heraclitus (difference) can be assigned as philosophical protagonists to a corresponding 'contemplation of identity' or 'contemplation of difference', and 'construction' (identity) and 'deconstruction' (difference) as basic operations of scientific theory. With G. Chr. Lichtenberg (1973) the 'shrewd' search for similarities in what is different, i.e., for something identical, can be compared to looking through a magnifying glass, which enables distance and an overview and thus provides insight into connections; the astute search for differences in what is similar, i.e., for something different, can be compared to looking through a magnifying glass, which enables proximity and thus provides insight into details. With Joachim Ritter, following Cusanus, a 'speculative' (from lat. *speculatio*, spying out, scouting) bird's eye view from a 'mirror tower' ('*cognitio speculativa*': 'speculative cognition'), which can see something "at once and as a whole at the same time", can be distinguished from a 'discursive cognition' ('*docta ignorantia*': 'learned ignorance'), with which "man follows the tracks across the field like a hunting dog" (Ritter 1974 [1963], p. 176). Gabriel assigns a specific worldview to each of the two 'cognitive faculties': "The analogical and aesthetic worldview manifests itself in the cognitive faculty of wit, the logical and scientific worldview in the cognitive faculty of ingenuity" (Gabriel 1996, p. 2). In view of this reconstruction, wit, humor, and irony can be defined in more detail regarding their epistemological and epistemic critiquing function.

In the case of *wit*, this function evidently lies in the "ability to come up with surprising ideas", i.e., in "shrewdness" (Willer 2017, p. 11). This shrewdness is characterized by finding something similar, connecting or unifying in that which is supposedly heterogeneous, unconnected, incompatible, or different. It was not until the nineteenth century that the word 'wit' was replaced by the "borrowed word 'genius', modeled on the Latin '*ingenium*' [...] If one originally had wit, one now had genius, without therefore already being a genius" (Gabriel 1996, p. 8). In early Romanticism, namely through F. Schlegel, wit was exaggerated to such an extent for art and poetry as jocular wit that wit displaced ingenuity and a 'universal poetry' abolished the boundaries between art and science or between poetry and philosophy in a 'unified science': "Everything is wit and everywhere is wit" (Schlegel 1981, p. 153). In this way, the scientific view of the world is "incorporated into the aesthetic" (Gabriel 1996, p. 11). Conversely, the scientific view of the world could attempt to make logical ingenuity absolute with its conceptual work and its conceptual boundaries and to incorporate the aesthetic view of the world within the framework of a scientific 'unified science'—as was sought, for example, in logical empiricism (Kühne und Berr 2021). The epistemological function of jokes also remains intact in their comic use, insofar as the comic effect of jokes can be attributed to the fact that "heterogeneous things are placed in an unexpected and thus laughter-inducing relationship" (Gabriel 1996, p. 12).

Accordingly, *humor* consists of the "capacity for a form of 'epistemic self-distance' that results from a special kind of 'change of perspective'", i.e., a "form of self-relativization" (Kindt 2017a, p. 9). Humor is also described by many authors as an "attitude" or "virtue", which manifests itself in two ways: First, as the epistemological "insight into one's own limitations and therefore with a 'taking oneself back'" (Kindt 2017a, p. 8) and the other is as an attitude "based on a benevolent view of life and exemplified by a philanthropic form of comedy" (Kindt 2017a, p. 9). The first attitude can be seen in the fact that humor can basically be understood as the ability to detach oneself from a given context, to reflect on this context and not only to point out dissonances, but also to place them in a comic context. In this way, humor—and also irony—is a "means of communication, a form of expression that draws attention to something, to the unusual in the seemingly ordinary" (Grau 2010, p. 23). The second attitude manifests itself as a "warm-hearted, loving, benevolent and compassionate" ethos, "which is able to see the humanly stirring and endearing in the comic" (Hartmann 1953, p. 418). In this sense, H. Lützeler (1999) distinguished humor from wit. Jokes can reveal a hidden meaningful connection in a seemingly nonsensical context in a 'cunning' and intelligent way (Lützeler 1999), while in the sense of the epistemological tradition, uncover a similarity in difference. The punch line as an intensification ultimately illuminates the meaning in the supposed nonsense. It is a "game of the mind" (Lützeler 1999, p. 18) whose content, however, is of a very general nature and therefore detachable from concrete situations and circumstances. Humor, on the other hand, in the sense of a life-friendly, benevolent attitude, engages with concrete situations and circumstances and is bound to milieu and dialect, without claiming an essentialist nature of these dialects—Rhenish humor is different from humor in Hamburg, the Ruhr area, or that of the Swabian. This humor is life-affirming and environment-affirming and sympathetic to the 'creatureliness' and 'fragility' of people, whereas the joke is (positively) subversive, yet also (negatively) has an "exclusionary function by ridiculing origin, class and gender" Willer (2017, p. 14), etc. Humorous storytellers always include themselves in the life-affirming and compassionate aspect, the 'we' therefore plays a major role in humor, which in this sense represents a great social achievement in that it addresses and confirms the listeners in their own fragile humanity and conveys a certain 'warming' respectful acceptance and security. Such narrators do not take their idiosyncratic subjectivity as their starting point but turn to life and people as they actually are in the milieu-bound world. This is reminiscent of a well-known theorem of Hegelian aesthetics. Hegel finds, in Goethe's 'West-Eastern Divan', an example of a culture and worldview that approaches its world positively and with cheerful serenity. Hegel speaks of 'cheerfulness about things' and of 'objective humor' as opposed to 'subjective humor' (Hegel 2003 [1823] which revolves around one's own person and "one's own subjectivity, which produces itself" (Hegel 2003 [1823], p. 201) and not about the world, the worries, and needs of the specific listeners.

In *irony*, the "epistemic function of criticism" consists of "questioning something that is known as certain" (Wirth 2017, p. 17). In the literature on irony (as well as humor in some cases), a common thread of its interpretation can be reconstructed: It is always about gaining a distance from people, actions, ideas, and contexts; exaggeration and understatement of inauthentic (metaphorical) speech are typical; it is often about accepting the tension between ideals and profane reality; the 'subjective refraction of the real' plays an important role—and the inherent skepticism of supposed certainties and doubt about ultimate values, ideas, and orientations are often emphasized. The well-known 'Socratic irony' as maieutics ('midwifery') takes on the "habitus of a critical dialectic" and consists of Socrates confronting those "who believe they already know everything [...] not as the knower, but in the role of the ignorant and inquisitive" (Weinrich 2019, p. 578) in order to "shake the position of those who positively believe they have knowledge" (Wirth 2017, p. 17). The ultimate goal is, in Richard Rorty's vocabulary, to uncover, recognize, and acknowledge the contingency of language, the world, and human beings. Rorty also uses the Hegelian concept of 'dialectic', albeit in a sense related to Socratic irony. The 'romantic irony' already propagated by F. Schlegel also assumes in this sense that the ironic person can "rise above his limited life and exist in the metaphysical tension between the unconditional and the conditional, between ideal and reality"—it arises "from the feeling of finiteness and one's own limitation" (Weinrich 2019, p. 580). Schlegel therefore consistently defines irony as the "highest and purest skepticism" (cited in Vieweg 2005, p. 297), which as a "form of paradox [...] plays with the boundaries between 'jest' and 'seriousness'" (Wirth 2017, p. 18). Wirth (2017) compares this concept with what Luhmann calls 'frame confusion', which is the staging of "a model of transparent deceptions" (Luhmann 1995, p. 177). In Romantic poetry, the "abolition of all finitude and positivity of the finite through the infinite variety of subjective refraction of the real" (Preisendanz 2019, p. 1232) in other words, a transgression of boundaries in the sense of the analogous aesthetic conception of the world. Mann, for example, emphasizes the ironic distance. In his novels, he "often uses the role of the narrator to distance himself ironically from his own position and in this way achieves an 'ironic objectivism of epic'. For him, irony means distance, reservation, freedom" (Weinrich 2019, p. 581). These expressions (distance, reservation, freedom) implicitly connotate the concept of skepticism or doubt, as typical prerequisites for people who laughingly liberate themselves from constraints and dogmatism through wit, humor, irony, and satire. The Nobel Prize winner for literature Dario Fo expressed this as follows: “We are convinced that in laughter, in the grotesqueness of satire, lies the highest expression of doubt, the most important aid to reason” (quoted from Michaelis 1997, n.p.). This ‘help of reason’ is help for freedom from paternalism, impositions or purported alternative-free offers of orientation, namely as a “laughter that liberates itself from the orientation requirement as a reaction of recollection of itself, i.e., the rejection of the ultimate authority of orientation” (Gethmann-Siefert 2005, p. 186) as is demonstrated in an unsurpassable way in the comedies of Aristophanes.

Laughter, wit, humor, and irony can also be interpreted as forms of meta-reflection on the conditions of freedom against the background of what has been said so far (Kühne, Berr, Schuster, Jenal 2021). Freedom can be understood as ‘freedom *from*’ in the sense of liberation from something or becoming aware of something, or as ‘freedom *to*’ and ‘*for* something’ in the sense of voluntarily taking on views, tasks, obligations, commitments, and activities (Berlin 1995 [1969]). Lützeler, for example, limits himself in the chapter “Humor and Human Freedom” (Lützeler 1999) to ‘freedom from’. The “highest achievement of humor” is that people gain distance: They become “inwardly free, superior about the world and about themselves” (Lützeler 1999, p. 97). What remains unmentioned is the aspect of ‘freedom *to*’, which was already lacking in Romantic irony: the “sensorial act of laughter in the face of the dissolution of previous orientations deemed assured [...] is the act of liberation from the compulsion to orient oneself"—but not only that, it is “at the same time the act of compulsion to gain one’s own orientations. This last aspect, at least, is missing in romantic irony” (Gethmann-Siefert 2005, p. 183). Humor and irony are also about a “conscious enjoyment of finiteness”, because “in hilarity lies the positive attunement to a situation, the acceptance of the questioning of one’s own limits and the integration of one’s own limited lifeworld through other possibilities” (Gethmann- Siefert 2005, p. 185)—for example, the free engagement in a society *for something* or in science *for* a specific position of scientific theory.

2.3 The Theory of the Three Landscapes as an Analytical Approach

In order to facilitate access to the complexity of landscape understandings, the following explanations are based on the considerations of the analytical division of landscape into three levels (Koegst 2022; Kühne 2018b, 2021), which are derived from the three world theory of Karl Popper (Popper 1979; Popper und Eccles 1977). This approach can also be described as quasi-ontological (Berr et al. 2019). In the awareness that we have a causal dependency on the ‘world out there’, but no representational access (Rorty 1980, 1997[1989]), an ontological approach geared towards representing ‘the world out there’ can hardly be justified. However, since landscape research today is markedly based on ontological approaches to the world interpreted as landscape, whereas we follow the principle of the usefulness of (scientific) understanding of the world, we integrate ontological statements about the world in the form of an 'as-if' (cf. exemplarily: Vaihinger 1911) or quasi-ontology. In the concrete case of the theory of the three landscapes, this means that we are aware of the constructed character of the three landscapes, however, this theory has proven itself in various ways: First, with regard to the classification of theoretical approaches to landscape (and also space), second, with regard to the handling of complexity resulting from the object 'landscape', third, with regard to conceptual considerations of scientific work (such as the selection of methods; see Hinz et al. 2024; Kühne und Jenal 2020; Kühne und Koegst 2023b).

In his three world theory, Karl Popper differentiates between the physical World 1, the World 2 of individual consciousness, and the World 3 of cultural and socially shared content. World 2 occupies a central position here; every connection between World 3 and World 1 runs through World 2. As the bearer of World 2, the human being has a 'share' in all three worlds (to put it in quasi-ontological terms) and is therefore in a position to intervene materially in World 1. Through socialization, he appropriates 'parts' of World 3. In addition, he is able to have an innovative influence on World 3 by developing new interpretations, evaluations, and categorizations in World 2 and offering World 3 for integration. However, World 2 can also simply adopt and update the contents of World 3 (thereby confirming them). World 2 is not only integrated into World 3, but also into World 1 because of its physical constitution, to whose conditions the human being is subjected. As a result of his physical constitution, he is also able to observe World 1 (Fig. 2.1).

According to the theory of the three landscapes, a distinction can be made between a Landscape 3 (the social ideas, patterns of interpretation, categorization, and evaluation of landscape), a Landscape 2 (the individual ideas of landscape), and a Landscape 1 (the components of the physical world—i.e., World 1 for Popper—into which landscape is synthesized on the basis of Landscape 2 and Landscape 3 ideas). The constitutive level here is primarily Landscape 3, where the social stock of knowledge is stored, which serves as the basis for the creation of landscape, and secondarily Landscape 2, since without individual actualization the social stock of knowledge, Landscape 3, cannot become effective, but also because only Landscape 2 can have

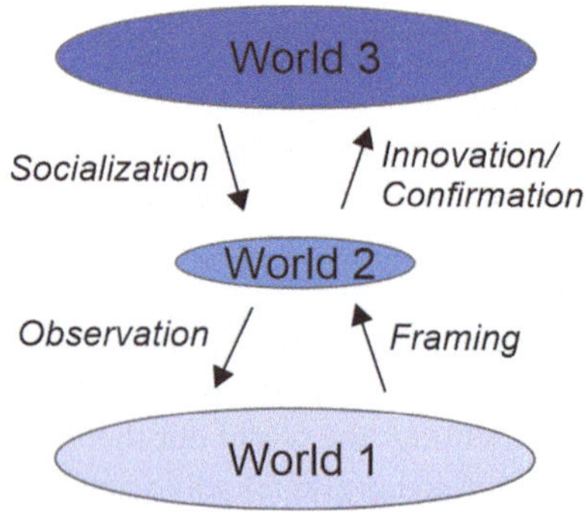

Fig. 2.1 The theory of three worlds according to Karl Popper (Figure based on: Kühne 2020)

an innovative effect on Landscape 3. Individuals integrate social ideas, patterns of interpretation, categorization, and evaluation of landscape, link them with individual experiences and give them meaning. They have a body and are thus in a position to have an effect in the physical World 1, with landscape interpretation, i.e., also in relation to Landscape 1.

On all three levels, landscape stands in a context to space (which can be constructed analogously as Space 1, Space 2, and Space 3). However, landscape is neither congruent with space (otherwise there would be no need for conceptual differentiation) nor is it merely a subset of space. The concept of landscape extends beyond that of space when it is used metaphorically, for example in the context of 'coalition landscape' (here the existence of and relationships between entities) or 'religious landscape' (in the sense of the distribution of religious affiliations). A central difference lies in the constitutive level; if landscape—as a social synthesis pattern—is (as already mentioned) constitutively bound to World 3, here as Landscape 3, the experience of spatial contexts begins quasi at birth (Piaget und Inhalder 1975 [1947]; Fig. 2.2).

The social content of landscape is subject to social and linguistic differentiation. Even within European languages, the understanding of 'landscape' differs considerably. What is understood by 'landscape' in German is only partially covered by the French 'paysage', which primarily encompasses the aesthetic construction of landscape. The German 'Landschaft' also includes the meanings of 'pays' (land, especially in political interpretation) and 'terroir' (area, surroundings). However, the meaning of Heimat, as it is present in German, is not found in French, but it is in Hungarian (Drexler 2009, 2013). What is outlined via the German word 'Landschaft' is present in numerous non-European languages only as a feudal word, such as in Turkish (*Peyzaj*) or Chinese (景观, *Jingguan*; Bruns und Münderlein 2019; Küchler und Wang 2009; Zhang et al. 2013). The social differentiation of the construction

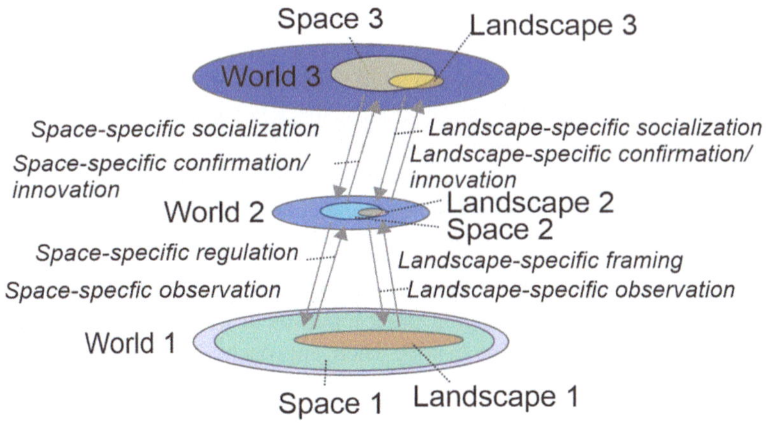

Fig. 2.2 The connections between the three worlds, spaces, and landscapes (Figure based on: Kühne 2020)

of landscape can be assigned to different modes of landscape socialization (Kühne 2020; for more detailed information on the basics, see Hokema 2015; Jenal 2019; Kühne 2018a; Peng 2020; Stotten 2013; Wojtkiewicz und Heiland 2012):

1. In Mode A, the 'native normal landscape' (German: 'Heimat') is created. With the mediation of significant others (such as parents), an affirmative emotional residence within the home environment takes place in early childhood, which is understood as instructive pertaining to what is 'landscape'. In normative terms, Landscape 1a, being formed in this way, is linked to stability, i.e., it is expected not to change. This mode of landscape construction is strongly influenced by the individual experience of World 1 as Landscape 1.
2. In Mode B, the *'common sense' understanding* of landscape is generated. Since this generally includes widely shared patterns of categorization, interpretation, and evaluation of landscape, Level 3 of landscape (Landscape 3) has a constitutive effect here, not the individual experience of World 1. The *'common sense' understanding* of landscape includes, notably, aesthetic and now also environmental content. These are formed and reinforced in school lessons, through television, through the internet, through various products of the press, etc., but also through negotiation processes with other people. Mode B also makes normative demands on Landscape 1, namely that it corresponds to the aesthetic and environmental stereotypes shared in the *'common sense' understanding.*
3. In Mode C, 'expert-like special knowledge' is generated. These Mode C landscape constructions are strongly characterized by disciplinary traditions, different theoretical approaches, and (quite possibly unreflected) ideological foundations. Mode C constructions of landscapes are strongly competitive, on the one hand with each other but also with Mode A and Mode B landscape constructions, from which they are distinctively set apart. The competitiveness between them produces an inherent pressure to innovate (only innovations can be published, justification of the need for third-party funded research, etc.). Norms are also associated with the Mode C construction of landscape. These norms are in turn subject to a consensus, for example in one's own discipline, in one's own research program, in one's own landscape-related worldview, in short, different vocabularies are formed. The approach to Landscape 1 is based on a view of deficits. This deficit view differs greatly between the individual vocabularies, resulting in a blatant contradiction of objectives between the individual vocabularies (e.g., between a conservationist, a historical conservationist, and an agricultural economist).

Referring back to the basic features of neopragmatism, it can again be stated that against the background of contingent interpretations of the world, Mode C cannot claim a prominent position from 'truth'. Their approaches to landscape must also prove useful for solving concrete problems. Richard Rorty (2018 [1994], p. 26–27) is able to recognize another function for science when he states: "Of Peirce's, Dewey's and Popper's praise of science [...] no specific epistemological strategy remains, but nothing more than a praise of certain moral virtues, namely those of an open society". In a nutshell, science can have the function of developing ideas that increase the openness of society.

2.4 The Relationship Between Dynamis and Energeia

Traditionally, following Aristotle's distinction between 'dynamis' (capacity, power, possibility) and 'energeia' (activity, reality), the distinction between potentiality (latin: potentia) and actuality (latin: actus) or between possibility and reality is used to define ontic structures. In this understanding, dynamis is the ability "to bring about a change in another object or in oneself" (Schwemmer 2004b, p. 516), energeia is "reality in the sense of a realized determination and possibility" (Gatzemeier 2004). Every movement (Greek: kínesis; an object changes location) and every change (Greek: metabolé; an object changes a specific property) is based on a "capacity for movement or change" (Schwemmer 2004a, S. 60). This capacity can be active as a principle of movement or passive as the ability to suffer something. The Aristotelian dynamis, whether active or passive, is not yet conceived logically as pertains to the three modes, but is to be distinguished as "*potentia realis*, i.e., as a possibility in reality or as something real in terms of its disposition [...] from *logical* or *objective* potency as mere conceivability" (Schlüter 2019 [1971], p. 135; emphasis in original), therefore, it is understood ontologically and teleologically as the principle of a purposeful realization, i.e., as 'entelechy'. In this sense, dynamis can also be defined by Aristotle as substance (Greek: hýle) and energeia as form (Greek: eidos, morphé). Aristotle therefore understands energeia as the reality "that the thing exists, not in the sense in which it is said to exist according to possibility—for according to possibility we say, for example, that there is a Hermes in wood and in the whole line its half, because it could be removed from it, and we also call a scientist according to possibility the one who is not engaged in theoretical knowledge, but who is only capable of generating it—but according to actual activity. [...] For just as the builder relates to the builder-artist, so also the waking relates to the sleeping, the seeing to that which closes its eyes but still has the sense of sight, that which is separated from the material to the material, the processed to the unprocessed. In this contrast, the first member is intended to denote the actual, the other the possible" (Aristoteles 2009 [1st Century. BCE] Book 9, Chapter 6 ff.). In other words: "The reality of the possible, insofar as it is possible" is "movement" for Aristotle (Aristoteles 2009 [1st Century BCE], Book 11, Chapter 9). The "reality of the possible" therefore manifests itself in an act that realizes that which is possible. In the example of the statue: the stone is considered a potential statue. It is then about the transition between the two spheres. Everything that exists has these two sides: possibility and reality. It depends on the movement to transform the 'possible' into the real.

In medieval scholasticism, namely by Thomas Aquinas, "categories for the conceptual representation of movements, changes and becoming" were developed with actuality and potentiality (Schwemmer 2004a, p. 60). In this epoch, they acquired "a fundamental function. However, very different philosophical approaches are concealed under a rather uniform terminology" (Schlüter 2019 [1971], p. 134) and philosophical developments across history, which cannot be discussed in more detail here. What is decisive for the further argumentation in this book is the indication that the talk of contingency and inversion must not be misinterpreted ontologically or teleologically, but that both categories denote possibilities of thought and corresponding possibilities of action that affect (energeia, actus) physical landscapes (World 1) through individual and group-specific actions (World 2), mediated by social and cultural objectivations (World 3).

Throughout the nineteenth and twentieth centuries, the ancient, medieval, and modern ontological or teleological conception of reality was gradually countered by a conception of possibility that confronts the often essentialist conception of reality with the "category of the possible" as a corrective. This process has now progressed so far that in science, philosophy, art, and aesthetics, that which Musil has (1992 [1930]) referred to as the "sense of possibility before the sense of reality" (Gamm 2011, p. 30) takes precedence. Even in the sciences, 'reality' must be tested in elaborate experimental systems (cf. Knorr-Cetina 2002; Latour und Woolgar 2013 [1979]; Rheinberger 2001) and prove itself pragmatically in its assumed veracity. In philosophy, in the sciences, as well as in everyday life, the possible, the hypothetical, the fallible, the provisional, the experimental, the trial and draft, the reality in the subjunctive, as it were, now comes to the fore. In the philosophy and theory of science, paradigm shifts and research programs, heuristics and hypothesis falsifiability, 'evolutionary world views' (Poser 2012) and the recognition of uncertainty in the context of 'mode 2 knowledge production' (cfGibbons et al. 1994; Kühne und Berr 2021; Nowotny et al. 2001; Prominski 2004), thus emphasizing the draft-like, the trial and experimental. In each case, the reflective playing through of possibilities is also a decisive factor in the acquisition of knowledge and in world orientation, so that one can even speak of an 'art of the possible' (Hubig 2006, 2007, 2015).

What is decisive in this context for our further considerations is the transition from a non- thematically given possibility (contingency) in a space of the conceivable to a thematization of a specific conceivable possibility, which, although not yet 'real', can be specifically anticipated in its 'real' consequences and secondary consequences in thought (or virtually or experimentally). In this respect, this thematization, which we will refer to as 'inverse' (see Sect. 3.1), represents a tipping point between mere contingency (dynamis, potentia) on the one hand and (entelechially conceived) realization or (modally and logically conceived) reality (energeia, actus) on the other—an intermediate figure that was already well known to the conceptually highly differentiated scholastic tradition (cf. Schlüter 2019 [1971]).

2.5 Neopragmatism in Reconnection—Psychological and Communication Considerations

In the lifeworld,[1] this tipping point from possibility to realization is embedded in a complex system of situational, personal, and social interaction conditions. In the process of perception, contingency must first be perceived as a range of options, i.e., as personal room for maneuver. The structure of ambiguity in turn results in a field of perception that is additionally processed with associations and cognitive processing mechanisms. To analyze these tipping points, an examination of perceptual and cognitive psychological processes involved in such constructions of the world is therefore apparent.

From a psychological point of view, it is notable that when contingencies are perceived, the desire arises to initiate a process of conclusive recognition, which is linked to the hope of being able to bring it to an end ("need for completion" according to Lewin 1927, p. 85) This tendency towards a 'closed form' is embedded in an intuitively holistic (top-down) and analytical step-by-step (bottom-up) processes of perception, in which the new and contingent is compared with what has already been experienced (Schuster 2021). Depending on the environmental requirements, this process can take place automatically or with directed attention. Perceptions are fundamentally shaped by individual cognitive processing and the contextual conditions of perception. Perception and cognitive processing merge into one another and are mutually dependent as soon as stimuli are beyond the threshold and reach the level of perception (cf. Birbaumer und Schmidt 2018). In science, this is demonstrated by the fact that concepts from the psychology of perception (top-down and bottom- up processing) appear in a similar way in cognitive psychology (e.g., Kahneman 2014). The close connection between perception and cognition indicates that cognitions do not lead to the mapping of environments, but function as tools to deal with them (Müller 2014). This means that there is no *objective* perception, but that every perception is comprised of mental processing, supra-individual and individual embeddings, and the resulting expectations (e.g., Ansorge und Leder 2017; Becker- Carus und Wendt 2017; Goldstein und Cacciamani 2023; Schönhammer 2013; Treisman 1993.

The special significance of the interaction of expectations and the objective world of experience in the context of perceptual processes is addressed by the ecological approach to perception according to Gibson (1982) in which he uses the concept of affordances to attribute a special role to the perceived offers of action or the prompting character of objects. According to this concept, objects and environmental conditions in themselves contain information on how they can or could be handled. The physiological components of the informational processing of individual stimuli in the context of perceptions are of secondary importance in this conception (cf. Gibson 1982, p. 40).

Rather, the focus is on the interwovenness of perception and expectations, through which attention is directed in such a way that the threshold of perception is raised

[1] Lifeworld: all the immediate experiences, activities, and contacts that make up the world of an individual or corporate life.

for certain target stimuli and lowered for all others through cognitive focusing. In this perspective too, perceptions are not predominantly logical as well as analytical processes of cognition, but the byproduct of our attention, wishes, expectations, and feelings. Emotions in turn regulate the energization in the process of perception and act as a kind of "relevance detector" (Eder und Brosch 2017, p. 195) of attention orientation.

Modeling factors are personal relevance, the assessment of consequences, and personal coping potential as well as the evaluation of the situation with regard to one's own values, attitudes, and perceived social norms (Scherer 2009). As regards contingencies, it can therefore be assumed that the motivation to want to deal with them depends on the fit with the perceiver's personal lifeworld. In lifeworld terms, this means that the desire to engage in-depth with different 'explanations of the world' is not equally distributed, but varies depending on how it is embedded in the personal semantic network (e.g., Waldmann 2018). However, the close relationship between semantic networks and language and culture creates supra-individual similarities, so that there are often socially shared and group-related associative links.

Interest in, and prior experience with, certain perceptual situations also systematically interfere with the motivational system of wanting to deal with contingencies. In the context of landscape perception, for example, prior experience and interest lead to a higher appreciation of complex landscapes (Kühne 2008) which for architecture (Rambow 2000; Schuster und Kirschbaum 2011; Stamps und Nasar 1997) and art (Raab 1981) has also been empirically proven. Mehrabian (1987) explains this connection between perceived formal and structural stimulus qualities and the cognitive processes of understanding, i.e., the assignment of content to a structure of meaning and knowledge, as follows: "If we have succeeded in giving meaning or structure to a complex environment, we have not only reduced its complexity, but we have also made it more familiar or less improbable, thus dampening its overall volume of stimuli" (Mehrabian 1987, p. 19).

To summarize, the perceptual and cognitive psychological observations lead to the conclusion that when people perceive contingencies, they feel the desire to deal with the ambiguities, but at the same time there is a reflexive tendency to shoot quickly. This means that intuitive and automatic forms of thinking (heuristics) are given preference over the more complex reflexive and rational processes of thinking (Döring 2022; Gigerenzer 2014; Kahneman 2014). One phenomenon—and to some extent problem—of heuristics is that they are less time-consuming because of their simplicity and are therefore perceived as less strenuous than in-depth thought processes. The fundamental attractiveness of 'cognitive shortcuts' means that quickly finding a (final) solution is the motivational norm—even for people who have the potential to operate in a more elaborate way. In the cognitive science literature, this is partly why people are referred to in general terms as "cognitive misers" (Fiske und Taylor 2020; Spears und Tausch 2014, p. 522). In combination with the tendency to select information to confirm already held opinions and expectations (Nickerson 1998), an open and directionless confrontation with contingencies is not the obvious spontaneous reaction.

Overall, as suggested by cognitive psychology analyses, each person tends to master their own vocabulary for describing the world and wants to protect it from disruption. In the process of the debate on the 'better version of the world description', two tendencies are therefore obvious, both of which spontaneously go hand in hand with a low depth of debate and lead to encapsulation towards new perspectives: 'Affection' when the other vocabularies agree and 'rejection' when they diverge. Rorty calls the latter: "alternately distrustful" communities of justification (Rorty 2023, p. 178). The term 'communities' implies that there are linguistic and cultural socialization processes that produce acceptance and rejection groups. To break through the tendency of self-affirmation and encapsulation, Rorty refers to a self-referential basic orientation of irony—an attitude intended to strengthen openness, change of perspective, and tolerance of ambiguity (not least through humor). However, since it cannot be assumed that all interacting partners are able or willing to always demonstrate the necessary degree of maturity in all situations of interaction, care must be taken on the sending side to ensure that the motivation to deal with contingencies is strengthened. Since simplification and the tendency towards coherence is the cognitive norm that is not easily abandoned, communicative forms of uncertainty are the obvious way to encourage people to leave the comfort zone of a closed description of the world. Techniques of questioning with the aim of dissolving or at least loosening already closed vocabularies are helpful for discourse (Schwerz-Friesel et al. 2012). Whatever technique is used for this (irony, criticism, etc.), it must be chosen in the sense of the neopragmatic approach in such a way that it can lead to the "expansion of the 'we'".

Discrediting forms of communication, such as provocation, insults, or attacks, are not suitable for this purpose. Rather, it is important to provide suitable motivational support for the potentially aversive state of dissolving an established point of view. In communication processes relating to contingent descriptions of the world, it is not only the content of intellectual stimulus that needs to be considered, but also the motivational dimension of wanting to open up and make productive use of the phase of uncertainty. In principle, the method of questioning and unsettling requires a communicative attitude that is based on a separation of content and sender, i.e., the factual aspect, self-disclosure, and relationship side are considered separately (see Schulz von Thun 2019) to be able to dissolve the link between person and theory. According to Rogers (2005), the three characteristics of growth-promoting communication—empathy, congruence, and emotional appreciation of the other person—should be considered in order to create a constructive culture of conversation. In communicative practice, mutual appreciation is particularly evident and develops through the inner attitude of giving and receiving, which "…is of the greatest importance in purely human terms" (Popper 1996, p. 160).

2.6 Neopragmatism in Perspective—Life Chances and Symbolic Capital

The central normative element of neopragmatism is—as shown—the possibility of private self- creation (Rorty 1997 [1989], 2023). The possibility of the unfolding and development of World 2 only occurs partly out of itself; rather, it is also dependent on other Worlds 2, especially with regard to the accessibility and ability to socialize the contents of World 3. Two concepts that shed light on this relationship are that of life chances by Ralf Dahrendorf (1979; see the introduction: Kühne 2017; Kühne und Leonardi 2020; Leonardi 2014) and that of symbolic capital by Pierre Bourdieu (Bourdieu 2012; see in more detail: Fuchs-Heinritz und König 2014; Jurt 2012).

2.6.1 The Concept of Life Chances and Its Expansion

Ralf Dahrendorf (2007, p. 44) sees life chances as "first and foremost opportunities for choice, options. They require two things: the right to participate and an offer of activities and goods to choose from". The opportunity to choose must be associated with meaning. Dahrendorf summarizes life chances (1979, p. 24) metaphorically as "the baking forms of human life in society; they determine how far people can develop". With his understanding of life chances, Dahrendorf expands on Max Weber's original understanding, which referred to socially predetermined possibilities for action. Life chances are dependent on social contexts: they are "opportunities for individual growth, the realization of abilities, desires and hopes, and these opportunities are provided by social conditions" (Dahrendorf 1979, p. 50). Even if Dahrendorf ascribes a certain degree of freedom to the individual in his concept of life chances, which is made possible by society, society also restricts life chances, as the example of class differences makes clear when "the possibilities of some (the 'ruled', the 'dependent') are decisively pre-structured by the decisions of others ('the rulers')" (Niedenzu 2001, p. 178). Dahrendorf's expansion of Weber's understanding of life chances was not least an outcome of the fact that he "adds a dimension of meaning to the concept; furthermore, from a theoretical perspective, it becomes clear that […] Dahrendorf assumes a socially structured elective behavior of social actors" (Mackert 2010, p. 413). Dahrendorf thus accomplishes a change in understanding towards a higher degree of activity (Kühne 2024). The differentiation of options and ligatures provides an additional expansion of understanding.Dahrendorf understands ligatures as "structurally predetermined fields of human action. Individuals are placed in ties or ligatures by virtue of their social positions and roles" (Dahrendorf 1979, p. 51). Ligatures produce obligation on the one hand, but also affiliation on the other, and constitute the "foundations of action" (Dahrendorf 1979, p. 51) such as "the ancestors, the homeland, the community, the church" (Dahrendorf 1979, p. 51). Ligatures bind the individual into society, while options as "choices, alternatives of action" given in "social structures" (Dahrendorf 1979, p. 50) and thus enable the individual to make choices that stem from social contexts. Options "are thus open for the future" (Dahrendorf 1979, p. 108). Clear classifications can be found with regard to the passive and active dimensions: "Ligatures are given, options are desired" (Dahrendorf 1979, p. 108). Only in interaction do life chances arise as "opportunities for individual action that result from the interrelation of options and ligatures" (Dahrendorf 1979, p. 55). Accordingly, no life chances can arise without the other; "Ligatures without options mean oppression, while options without ties are meaningless" (Dahrendorf 1979, p. 51–52). Ligatures differ in terms of whether the individual is at their mercy without being asked or whether they are entered into voluntarily, "that the family and the village are no longer communities of fate, but are increasingly becoming communities of choice" (Dahrendorf 1979, p. 52). Thus, ligatures engaged voluntarily make it possible to give meaning to individual options. For Dahrendorf, however, the expansion of life chances is "option politics, i.e., the struggle for the expansion of human life chances through the multiplication of options" (Dahrendorf 1983, p. 123–124).

In various empirical studies relating to life chances, however, it has become clear that both the understanding of options and that of ligatures require differentiation, which we have undertaken—via different configurations of researchers—and made fruitful for the interpretation of empirical results (in more detail: Berr und Kühne 2024; Hinz et al. 2024; Kühne et al. 2022; Kühne 2024; Kühne, Berr, Schuster, Jenal 2021; Kühne und Koegst 2023a). Options can be divided into (a) latent options that are available to society as a whole (such as the right to freedom of expression), (b) individually available latent options, as the subset of latent options that are actually available to a person (such as having access to social contexts, expressing one's own opinion), (c) manifest options, i.e., when a latent option is actually taken (for example, when the person decides to actually express their own opinion). Beyond the latent options, there are also potential options that are not yet (latently) available under current social conditions (Fig. 2.3).

The relationships are not the same for all individuals, i.e., within the set of latent options the individually available options can be quite different, the intersection can be small, but they always remain present (thus a certain set of individually available options is guaranteed by the fundamental rights). Accordingly, two different individuals can manifest completely different options (Fig. 2.4).

Socially, but also individually, the scope of the different types of options is not stable, instead—to a certain extent—variable: socially, latent options can be expanded into the realm of potential options (for example, through technical aids), the individually available options can be expanded (for example, through education) and by increasing individual skills (such as improved time management) the number of individually manifested options can also increase (Fig. 2.5). Options of different types can also shrink, which can be particularly attributed to ligatures—which will be discussed below.

The differentiation of the understanding of ligatures is much more diverse. First, it emerges from the distinction between voluntary and involuntary made by Dahrendorf and mentioned above (Dahrendorf 1979). Second, moral ligatures bind the person into a structure of unquestioned norms. Ethical ligatures, on the other hand, give the

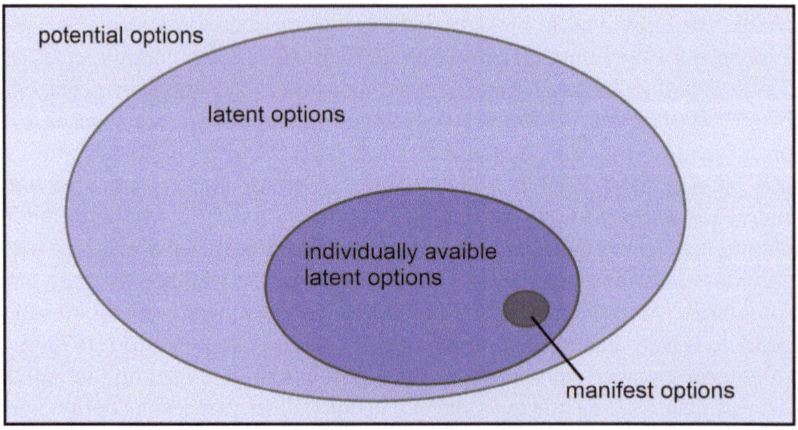

Fig. 2.3 The relationship between latent options, individually available latent options, and manifest options, as well as potential options as a representation of quantity (graphic: Olaf Kühne)

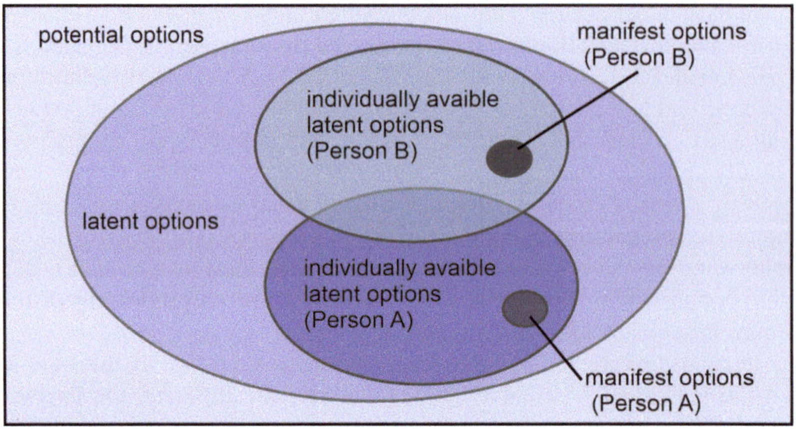

Fig. 2.4 Different individually available latent options and the resulting different manifest options of a second individual (Graphic: Olaf Kühne)

person the opportunity to reflect on options and ligatures (including moral ones). Moral ligatures can be differentiated, in turn, into those of compassion and those of resentment; the former result in dealing with others by means of empathy, the latter by means of rejection and devaluation. Third, inwardly directed ligatures are directed towards oneself. Outward-directed ligatures, conversely, are directed at others. The degree of externalization varies from specific (only one other person can be affected) to general (in extreme cases, all of humanity). Fourth, explicit ligatures are identified as ligatures, implicit ligatures are assumed to be known without being made clear. Fifth, physical ligatures exist as a result of the physical limitations of human beings and their mental attachment to this body. Physically mediated social ligatures, in

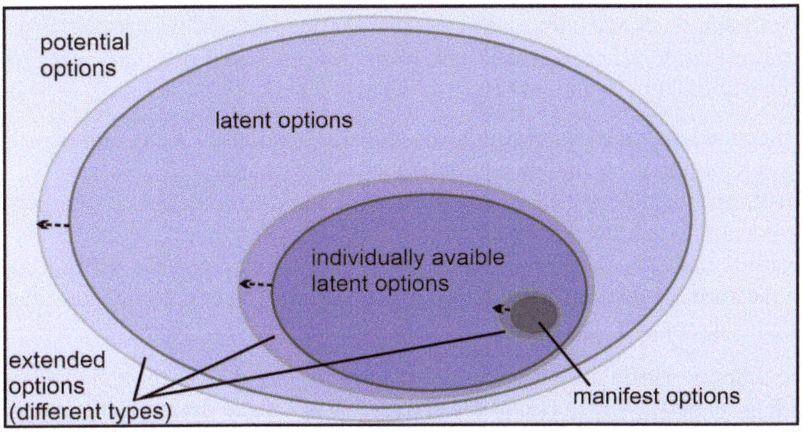

Fig. 2.5 Extensions of the options (of the different types; graphic: Olaf Kühne)

turn, arise through the bodily ligatures when social action is directed, but also made possible, through the 'being-in-the-world' of World 1 of humans.

We will return to this expanded concept of life chances on various occasions for an analytical basis when it comes to understanding how different ligatures and options play out in the context of inverse landscapes. On the other hand, we follow the normative dimension of society to expand life chances.

2.6.2 Symbolic Capital and Symbolic Power Including Their Spatial Extension

Bourdieu's concept of symbolic power (e.g., 1985b; Bourdieu 1985a und 1987 [1979]), as well as Bourdieu and Passeron (1973), deals with the ramified relationships of power in society. Bourdieu and Passeron (1973, p. 12) understand symbolic power (or symbolic violence) as "any power that succeeds in imposing meanings and asserting them as legitimate by concealing the power relationships that underlie its force". Power is therefore closely linked to giving one's own interpretations and evaluations (temporary) binding force. This results (Bourdieu (1987 [1979]) from variated access to symbolic capital. This can be understood as opportunities to gain and maintain social recognition and social prestige in society, whether in society as a whole or in parts of it (Bourdieu 1987 [1979]). Symbolic capital is formed from economic, social, and cultural capital. What all forms of capital have in common is their scarcity and desirability (Bourdieu 1985b and 1987 [1979]). Economic capital can be understood as material possessions that can be exchanged for money. Social capital "is described as a relational good inherent in social relationships and as a resource of different social structures with different social scope for individuals and corporate actors or communities, of various social structures with different social reach for individuals and corporate actors or communities" (Maischatz 2010, p. 31). Cultural capital does not appear in one form, but is divided into three sub groups (Bourdieu 2005 [1983]; Jurt 2012):

a) Objectified cultural capital comprises physical manifestations of human activity (books, buildings, technical equipment, works of art, etc.).
b) Incorporated cultural capital, on the other hand, is directly linked to a person's physicality (such as education or cultural skills).
c) Institutionalized cultural capital represents parts of incorporated cultural capital in the form of educational titles (e.g., master craftsman's certificate, master's degree, doctorate).

The different availability of symbolic capital results in a stratigraphy of society (as well as its parts), which (Bourdieu (1987 [1979]) can be divided into three basic classes:

1. The ruling class is made up of entrepreneurs, with a high level of economic capital but a low level of cultural capital and intellectuals, who have an inverse relationship between cultural and economic capital. The ruling class is the bearer of legitimate taste. This is characterized by a 'sense of distinction'. This taste shapes (especially aesthetic) points of view whose meaning and significance (initially) elude the other classes.
2. The middle class (also known as the petty bourgeoisie) is the bearer of middle-class taste. This is characterized by 'educational affinity'. Associated with this are constantly repeated attempts to appropriate the patterns of interpretation and evaluation, which, however, fail because of the lack of incorporated cultural capital.
3. The dominated class in turn, is the bearer of popular taste. This is oriented towards 'the decision of what is necessary'. This 'dominated class', with its modest endowment of symbolic capital, forms the rest of society.

This structure of society, in which "the subjective intentions and the individual or collective designs" (Bourdieu 1979 [1972], p. 179) are shaped by the respective class interests, is secured by mechanisms of persistence, especially in that the ruling class as well as the middle class strive to secure their stock of symbolic capital vis-à-vis the class below them through differentiation. With regard to social capital, the safeguarding of the stock is achieved in particular through social closure (Maischatz 2010). Incorporated cultural capital, for example, arises through the cultivation of time-consuming skills to be learned; as regards economic capital, for example, through high economic access costs to certain activities (such as sailing or golf).

Even if the class structure interpreted in this way by Bourdieu was increasingly overlaid by a society characterized by diverse and, above all, horizontally described forms of milieus and lifestyles (Hradil 2006, 2018; Schulze 2000), Bourdieu's patterns of the use of symbolic violence to secure dominance by means of symbolic capital through the distinctive devaluation of alternative world interpretations and value systems can also be found within certain sub-societies—and between them (in relation to landscape: Kühne et al. 2022; Kühne, Berr, Schuster, Jenal 2021).

Another differentiation of symbolic capital can also be found in relation to 'Space 1'. This differentiation can be found in Kühne and Koegst (2023a) as a social consequence of land loss occurring along the Louisiana coast. Indigenous populations are markedly affected by this loss of land, as they were displaced to this coastal region during the colonization of Louisiana by whites. There they have developed skills to adapt to the inherent ecological conditions, especially by using the waters. During land loss, not only do the ecological conditions change (such as the intrusion of salt water), but local communities also become unsustainable. The result is resettlement to more protected locations (Baum 2021; Burley 2010; Colten 2017, 2021; Hinz et al. 2024; Lambert et al. 2021; Weber 2024). Resulting from the centuries- long colonization of the hybrid space between sea and land (Kühne und Koegst 2022) they have thus built up a specific incorporated cultural and social capital that is only useful in a specific Space 1. Resettlement to another Space 1 now devalues this specific incorporated cultural capital as well as the social capital (which was formed, for example, in the context of fishing in dangerous waters). This makes it clear that the unequal distribution of symbolic capital does not only cause a stratification of society as an entirety, but that certain forms of cultural and social capital are (or can be) developed in certain Spaces 1.

According to Bourdieu's theory, resettlement does not mean a light-footed lateral movement into a different performance of the social habitus but requires active appropriation processes that have an arduous character, since the habitus is deeply rooted in the person through socialization (social origin and education; cf. Bourdieu 1987, pp. 17–18). Both the spaces and a suitable milieu-related material embedding plus active environment (Landscape 1) must be present and recognized. If the environment does not offer any options for activity that can be ascribed to the habitus, then it is difficult to appropriate the new environment. According to Ogburn, appropriation processes of material goods in immaterial culture are not always synchronous ("When material conditions change, changes are occasioned in the adaptive culture. But these changes in the adaptive culture do not synchronize exactly with the change in the material culture";Ogburn 1922, p. 203), instead, a kind of socio-cultural lag emerges ("...the material conditions change and the changes in the adaptive culture follow later"; Ogburn 1922, p. 211).

2.7 (Neopragmatist) Irony and Ironic Capacity: Concepts of Symbolic Capital and Life Chances

The figure of the 'liberal ironist' conceived by Rorty is characterized in particular by an attitude of "never quite being able to take themselves seriously, because they are always aware that the terms in which they describe themselves are subject to change" (Rorty 1989, p. 128). This irony is therefore characterized by the awareness of the contingency of language, self, and society as well as the lapse of conclusive vocabularies. These contingencies and this inconclusiveness of vocabularies, which are associated with a basic ironic attitude, also fundamentally mean "separation, restriction, liberty" (Weinrich 2019 p. 581) vis-à-vis supposed necessities, lack of alternatives, and ultimate foundations of justification. Expressions such as 'distance, reservation, freedom' connotate the concepts of skepticism and doubt, which can be seen as typical prerequisites for freeing oneself from constraints and dogmatisms through wit, humor, irony and satire, smiling, laughing, and amusing oneself—namely from paternalism, impositions, and supposedly alternative-free offers of orientation. In this respect, irony also manifests itself as a "laughter that liberates oneself from the orientation requirement as a reaction to a return to oneself, i.e., the rejection of the ultimate authority of orientation" (Gethmann-Siefert 2005, p. 186). Regarding such vocabularies, this means, in Rorty's sense, that the 'sovereign play with vocabularies' requires not only a thorough knowledge of different vocabularies, but also the rules of a (self-)ironic distancing in contrast to "common sense. For it is the watchword of those who unbiasedly describe everything important in terms of the final vocabulary to which they and their environment are accustomed" (Rorty 1989, p. 128).

However, this understanding of the individual and social opportunities of irony presupposes the ability of the addresser ('EGO') to use irony and the ability of the addressee ('ALTER') to understand irony. These abilities are of course themselves contingent, both on the part of the addresser and the addressee. Following Luhmann, we can therefore speak in this context of a 'double contingency' of EGO[2] and ALTER as regards the capacity for irony. However, it is not only these abilities of the addresser and the addressee that are contingent, but also the respective knowledge of the possibilities of these abilities or inabilities. If these abilities and this knowledge are simply assumed and no consideration is given to the fact that they may not be available, this can lead to considerable and serious misunderstandings, disruptions, and even the breakdown of communications and interactions. Instead of enabling freedom, irony can then also lead to reactance.

First, a problem of ironic speech is the proximity of irony to forms of comedy, which is always in danger of slipping into the ridiculous or mocking, or of being confused with and perceived as such forms, thereby also having an offensive effect (see Sect. 2.2). It is no coincidence that Schlegel (1967), with regard to 'romantic irony', spoke of a 'form of paradox' "that plays with the boundaries between 'jest' and 'seriousness'" (Wirth 2017, p. 18). Irony is therefore also a specific playful form, the rules of which must be known to everyone to enable fair and purposeful communication between 'EGO' and 'ALTER'. This playful competence in irony, i.e., the fundamental ability to ironize, is therefore, in Bourdieu's sense, dependent on symbolic capital (Bourdieu 2005 [1983], 2016; see Sect. 2.6.2) in the sense of a symbolic knowledge of the rules and limits of irony.

[2] Niklas Luhmann: dynamic social system interactions, double contingency, mutual influence, and circularity: EGO: An individual or entity selecting and communicating information to ALTERALTER: An individual or entity receiving this information from EGO.

Second, irony is characterized by an ambivalence in the social sphere. In the sense of Pierre Bourdieu's theory of capital and distinction (2016) irony can be used not only (positively) as a means of sensitizing contingency and opening up spaces of freedom, but also (negatively) as a means for speakers to gain distinction. In the sense of Ralf Dahrendorf's life chances approach (1979, 1992) irony can not only be (positive) options (see Sect. 2.6.1), but can also (negatively) restrict options and (positively) repel ligatures as repressive bonds (Kühne et al. 2022; Kühne, Berr, Jenal, Schuster 2021; Kühne und Koegst 2023a), but also (negatively) belittle meaningful bonds.

Bourdieu's theory of capital and theory of distinction show that the capacity for irony is dependent on individually accumulated *cultural* capital, because irony, insofar as it relates to specific worlds, is based on factual knowledge. In Aristotle's sense, the ironist must know both the 'center of truth' in order to be informed about factual criteria for determining this 'center', but an ironist also needs this factual knowledge in order to be able to determine the corresponding degree of deviation from the 'too little' or 'too much' that is accepted or rejected in certain social contexts. The ironist also needs a sound knowledge of people and society, as there must be an ability to assess from the habitus of the interlocutors whether and to what extent irony can be understood as a sign of self-contingency and not as a manifestation of a pejorative distinction. In other words, the ironist must also have *social* capital to understand and appreciate irony. In view of the aforementioned 'double contingency', this applies equally to the addresser and the addressee. In addition to cultural and social capital, economic capital is also required for the ability to use irony. For this ability must be acquired at a distance from society, language, and the self that can sufficiently detach itself from an existentially enforced perspective of an economic participant in the daily struggle to secure the fundamental needs of life. A tile setter, or a cashier, is unlikely to have the necessary freedom and time to acquire the sometimes subtle knowledge plus the cultural and social capital required to acquire a sufficient capacity for irony. This applies even more to the ability to 'play confidently with vocabularies', because this game not only requires a sound knowledge of many different vocabularies, but also a thorough and determined knowledge of the rules of ironic and self-ironic distancing in specific milieus, areas of life, and communities of discourse. However, this combination of specific skills can only be acquired by investing considerable time and resources. People who want to acquire these skills must potentially be able to devote time to the existential necessity of profane gainful employment to secure their personal livelihood. If this is not possible, not only will the acquisition of irony skills be restricted, but life opportunities will also be limited.

With Dahrendorf, the ambivalence of irony can be combined with the ambivalence of Dahrendorf's ligatures (see Sect. 2.6.1; cf. Kühne et al. 2024, p. 24–30).

- If irony arises from an awareness of the contingency of the self, language, and society, then it is incompatible with internal and external *moral* ligatures, as they lack the inner distance to norms that is constitutive of contingency sensitivity.
- Irony *can*, however, arise from *ethical* ligatures, as the adoption of an ethical meta-perspective on moral norms, their functioning and effect is a favorable, albeit

insufficient condition for the ability to use irony. Ethical reflections on moral ligatures can also take place without the creative and often playful character of irony.

- If irony is directed *inwards* towards oneself as an expression of ethical ligatures, it can contribute to individual self-coping, linguistic coping, and social coping, primarily concerning ligatures that are normatively imposed on the individual from *outside*. If ethical ligatures are directed *outwards*, they can bring with them considerable potential for conflict, especially in confrontation with other ethical or moral ligatures. Since ethical ligatures are usually accompanied by an awareness of the perspective dependency of one's own worldview and interpretation of the world and thus also tolerance towards other worldviews and interpretations of the world, the probability that the justification of ethical ligatures other than one's own will be recognized is not insignificant.
- In this context, irony reaches its limits when externally directed ethical ligatures are addressed to people in a specific context in which the ability or willingness to reflect ethically on morality is not present and ethical ligatures are potentially rejected. This can lead to the ironic reflection upon, and criticism of a moral 'obligation' associated with ethical ligatures being perceived as a normative imposition and not as critical reflection, but as an obligation itself. There is a risk of hierarchical communication in which mutual recognition of the legitimacy of other ties is no longer possible and the irony of ALTER can be interpreted as an attack, violation, paternalism or an attempt to gain distinction through EGO.
- To the extent that irony can play with *implicit* ligatures, this game is associated with ambivalence. On the one hand, this interplay in communication makes it possible to determine the boundaries of language games and to test whether a participant of communication belongs to their own language game or not and can therefore achieve solidarity or distinction. On the other hand, the practice of irony can also be used to distance oneself from explicit ligatures and thus either disavow the gain in freedom of explication or, through this explication, cast critical doubt on the previously implicit content.
- A major weakness of irony is bodily ligatures, which always convey the experience of the ineluctability of being-in-the-world of World 1, the vulnerability and the radical finitude of people in their World 1 boundedness. Although an ironic (self-)distance can also be cultivated to bodily ligatures, they can be excluded from conversations and discourses within certain limits, but they nonetheless exhibit a brutal commitment that can never be overcome and, for example, represents a merciless experience in the face of a serious illness.
- Physically mediated social ligatures are less implacable and can therefore be made the object of ironic references more easily than bodily ligatures. In this way, self-ironic reference can be made to World 1 changed by one's own interventions or ironically to objects shaped by other people. Ironical representations can also manifest a playful form of dealing with contingency (cf. the representations in Kühne et al. 2024).
- Bitter irony arises, for example, when access to cultural capital as the basis of the capacity for irony is denied and, because of the constraints of physical ligatures

(protection from the cold, obtaining food) in conjunction with the moral ligature of having to provide for oneself, individuals become victims of irony in two ways. First, resulting from a lack of symbolic and cultural capital, the capacity for irony is poorly developed; second, the formation of cultural and symbolic capital is hindered as an outcome of the compulsion to secure one's existence and therefore the lack of time resources and the compulsion to form economic capital.

In summary, the ability to deal with ethical ligatures also serves the acquisition of irony skills and a responsible approach to irony. This includes, first, not using irony as a medium for a distinctive devaluation of others, and second, not simply declassifying or radically deconstructing trust in the unambiguity and constancy of self, language, and society, but rather opening up the freedom- enabling perspectives and autonomy promoting effects of an ironic attitude towards self, language, and society. Understood in this way, irony can be a function of a worldview aimed at maximizing life chances.

2.8 Summary

With the theory of three landscapes, we have introduced a quasi-ontological framework. We therefore do not assume that there are three worlds in 'reality' but consider it a useful tool to classify the multiplicity of the world into these categories. Since we are causally bound to the 'world out there' (whereby our own World 2 is also not directly accessible in places and can therefore also be counted as part of the 'world out there'; see in more detail Freud 1955a [1917a], 1955b [1917b]), but cannot reproduce this representationally, only considerations of usefulness remain as quality criteria—also in connection with the fundamental considerations of pragmatism and neopragmatism. In this sense, the theory of the three landscapes derived from the three world theory has proven its worth in the past. Both represent tools for dealing with complexity and contingency. The aim is not to reduce complexity or limit contingency, but to be able to observe and develop complexity and contingency by means of a scheme for analysis. In Rorty's language, it represents a vocabulary (or part of one) that must constantly prove useful. Here, one of the three criteria of neopragmatism as a metatheory regarding the quality of (scientific) vocabulary formation becomes clear: achieving greater suitability compared to 'time-honored' vocabularies with regard to the interpretation of the world (in the context of landscape, for example, not making a distinction between landscape as an object and landscape as a construction). A second criterion for useful redescriptions is to be open to future redescriptions, for example, not to claim to have recognized an immovable 'essence' of something (the theory of the three landscapes fulfils this on the one hand through its self-description as a tool, and on the other hand through its departure from ontology). The third criterion is the aim to clarify contingency and the increasing of contingency (which, as shown above, is the claim of the theory of the three landscapes). Socially, the expansion of contingency can also be understood as the norm of inclusion, the expansion of 'we'. It is important not to design these vocabularies in such a way that they exclude others, but to include them with communication that promotes growth, i.e., empathy, congruence, and emotional appreciation. The fact that the process of expanding contingency is not without contradictions can be seen in two places: on the one hand, people are prompted by contingencies to deal with them, but also to reflexively achieve unambiguity by means of rapid firing. Contrastingly, in the neopragmatist sense, an ironic attitude means the development and application of a 'cheerful awareness of contingency'; however, this awareness of contingency is linked to the availability of cultural capital, which allows the option of playing with interpretations of the world. However, this cannot be assumed to be typically available. For a neopragmatist approach to the world, this means two things: first, avoiding the snubbing of people with a different level of cultural capital through the use of irony. Second, undertaking the task of supporting people in acquiring cultural capital, especially in the form of orientational knowledge. At this point, the potential of everyday language to engage as an interconnecting tool—even beyond the realm of everyday life—for translating different specialized languages becomes clear (Putnam 1997).

The associated generation and dissemination of ethical ligatures also serves to enable people to choose responsibly between existing options and to develop new options. Additionally, it serves to be able to question other ligatures, to be able to decide whether they should be engaged voluntarily or not. On a conceptual level, neopragmatism is operationalized with the life chances approach, both analytically and normatively. Analytically, especially in the form identified here, it makes it possible to examine individually, including supra-individually, shared and enabled options of various kinds, from potential to latent, and individually available latent options to individually manifested options. The same applies to ligatures, which can have a productive effect (such as the ethical ligatures mentioned above) or can also be aimed at restricting options (such as moral ligatures). In normative terms, the life chances approach is geared towards expanding them. In addition, the life chances approach is also given direction by neopragmatism. Life chances may otherwise remain individually directionless and guided by ligatures based on the search for options; with neopragmatism, they receive an orientation towards usefulness and inclusion, similarly. With the expansion of latent and individually available latent options, a transfer of a principally given possibility (contingency) into a space of possibility takes place.

References

Ansorge, U. & Leder, H. (2017). *Wahrnehmung und Aufmerksamkeit* (2. Aufl.). Wiesbaden: Springer.

Aristoteles. (2001 [ca. 335 – 322 v.u.Z.]. *Die Nikomachische Ethik. Griechisch-deutsch.* Düsseldorf: Artemis & Winkler (Übersetzt von Olof Gigon, neu herausgegeben von Rainer Nickel).

Aristoteles. (2009 [1. Jh. v.u.Z]. *Aristoteles' Metaphysik. Zweiter Halbband: Bücher VII-XIV* (4. Aufl.). Hamburg: Felix Meiner Verlag (Neubearbeitung der Übersetzung von Hermann Bonitz.

Mit Einleitung und Kommentar herausgegeben von Horst Seidel. Griechischer Text in der Edition von Wilhelm Christ.).

Barnes, T. J. (2023). Towards a pragmatist economic geography. *Environment and Planning A*. https://doi.org/10.1177/0308518X231203087

Baum, L. (2021). Umweltveränderungen, Klimawandel und sozialräumliche Folgen auf der Isle de Jean Charles – Eine kritische Analyse des medialen Diskurses. In O. Kühne, T. Sedelmeier & C. Jenal (Hrsg.), *Louisiana – mediengeographische Beiträge zu einer neopragmatischen Regionalen Geographie* (S. 151–166). Wiesbaden: Springer.

Becker-Carus, C. & Wendt, M. (2017). Wahrnehmung. In C. Becker-Carus & M. Wendt (Hrsg.), *Allgemeine Psychologie. Eine Einführung* (S. 75–176). Berlin: Springer.

Berlin, I. (1995 [1969]. Freiheit. Vier Versuche. Frankfurt (Main): Fischer.

Bernstein, R. J. (2008). Richard Rorty's Deep Humanism. *New Literary History 39* (1), 13–27. https://muse.jhu.edu/article/240141.

Bernstein, R. J. (2010). *The Pragmatic Turn*. Cambridge: Polity Press.

Berr, K. & Kühne, O. (2024). Landschaften sich schließender Gesellschaften. Eine Kritik am Leitfaden von Lebenschancen. In F. Bellin-Harder (Hrsg.), *Landschaft und Vegetation. Reflexionen zu Erkenntnistheorie, Herstellung und Interpretation* (Einzelveröffentlichungen Sozialwissenschaften, S. 147–180). Bielefeld: transcript.

Berr, K., Jenal, C., Kühne, O. & Weber, F. (2019). *Landschaftsgovernance. Ein Überblick zu Theorie und Praxis*. Wiesbaden: Springer VS.

Birbaumer, N. & Schmidt, R. F. (2018). *Biologische Psychologie* (7. Aufl.). Berlin: Springer.

Bourdieu, P. (1979 [frz. Original 1972]. *Entwurf einer Theorie der Praxis auf der ethnologischen Grundlage der kabylischen Gesellschaft*. Frankfurt (Main): Suhrkamp.

Bourdieu, P. (1985a). The Market of Symbolic Goods. *Poetics 14* (1-2), 13–44.

Bourdieu, P. (1985b). *Sozialer Raum und „Klassen". Leçon sur la leçon; 2 Vorlesungen*. Frankfurt (Main): Suhrkamp.

Bourdieu, P. (1987 [1979]. *Die feinen Unterschiede. Kritik der gesellschaftlichen Urteilskraft* (Suhrkamp Taschenbuch Wissenschaft, Bd. 658). Frankfurt (Main): Suhrkamp.

Bourdieu, P. (2005 [1983]. Ökonomisches Kapital – Kulturelles Kapital – Soziales Kapital. In P. Bourdieu (Hrsg.), *Die verborgenen Mechanismen der Macht* (S. 49–80). Hamburg: VSA.

Bourdieu, P. (2012). Ökonomisches Kapital, kulturelles Kapital, soziales Kapital. In U. Bauer, U. H. Bittlingmayer & A. Scherr (Hrsg.), *Handbuch Bildungs- und Erziehungssoziologie* (S. 228–242). Wiesbaden: Springer VS.

Bourdieu, P. (2016). *La distinction: Critique sociale du jugement* (Le Sens commun). Paris: Editions de Minuit.

Bourdieu, P. & Passeron, J.-C. (1973). *Grundlagen einer Theorie der symbolischen Gewalt*. Frankfurt (Main): Suhrkamp.

Bruns, D. & Münderlein, D. (2019). Interkulturelle Konstruktion. In O. Kühne, F. Weber, K. Berr & C. Jenal (Hrsg.), *Handbuch Landschaft* (S. 313–319). Wiesbaden: Springer VS.

Burley, D. M. (2010). *Losing Ground. Identity and Land Loss in Coastal Louisiana*. Jackson: University Press of Mississippi.

Buschmeier, M. (2023). Sprachphilosophie: Sprachspielpragmatismus und -historismus. In M. Müller (Hrsg.), *Handbuch Richard Rorty* (1st ed. 2023, S. 597–614). Wiesbaden: Springer Fachmedien.

Calder, G. (2007). *Rorty's politics of redescription* (Political philosophy now). Cardiff: University of Wales Press.

Colten, C. E. (2017). Environmental Management in Coastal Louisiana: A Historical Review. *Journal of Coastal Research 33* (3), 699–711. https://doi.org/10.2112/JCOASTRES-D-16-00008.1

Colten, C. E. (2021). *State of disaster. A historical geography of Louisiana's land loss crisis*. Baton Rouge: Louisiana State University Press.

Dahrendorf, R. (1979). *Lebenschancen. Anläufe zur sozialen und politischen Theorie* (Suhrkamp-Taschenbuch, Bd. 559). Frankfurt (Main): Suhrkamp.

Dahrendorf, R. (1983). *Die Chancen der Krise. Über die Zukunft des Liberalismus*. Stuttgart: Deutsche Verlags-Anstalt DVA.
Dahrendorf, R. (1992). *Der moderne soziale Konflikt. Essay zur Politik der Freiheit*. Stuttgart: Deutsche Verlags-Anstalt DVA.
Dahrendorf, R. (2007). *Auf der Suche nach einer neuen Ordnung. Vorlesungen zur Politik der Freiheit im 21. Jahrhundert* (Krupp-Vorlesungen zu Politik und Geschichte am Kulturwissenschaftlichen Institut im Wissenschaftszentrum Nordrhein-Westfalen, Bd. 3, 4. Aufl.). München: C.H. Beck.
Davidson, D. (2004). *Subjektiv, intersubjektiv, objektiv*. Frankfurt (Main): Suhrkamp.
Deines, S. (2008). *Situierte Kritik. Modelle kritischer Praxis in Hermeneutik, Poststrukturalismus und Neopragmatismus*. Frankfurt (Main): Johann Wolfgang Goethe-Universität.
Döring, T. (2022). *Öffentliche Finanzen und Verhaltensökonomik. Zur Psychologie der budgetwirksamen Staatstätigkeit* (2. aktualisierte und erweiterte Auflage). Wiesbaden: Springer Gabler.
Drexler, D. (2009). *Landschaft und Landschaftswahrnehmung: Untersuchung des kulturhistorischen Bedeutungswandels von Landschaft anhand eines Vergleichs von England, Frankreich, Deutschland und Ungarn*. Dissertation, Technische Universität München. München.
Drexler, D. (2013). Landscape, Paysage, Landschaft, Táj. The Cultural Background of Landscape Perceptions in England, France, Germany, and Hungary. *Journal of Ecological Anthropology 16* (1), 85–96. https://doi.org/10.5038/2162-4593.16.1.7
Eder, A. & Brosch, T. (2017). Emotion. In J. Müsseler & M. Rieger (Hrsg.), *Allgemeine Psychologie* (3. vollständig überarbeitete, aktualisierte und erweiterte Auflage, S. 185–222). Berlin, Heidelberg: Springer.
Fiske, S. T. & Taylor, S. E. (2020). Social Cognition evolves: Illustrations from our work on Intergroup Bias and on Healthy Adaptation. *Psicothema 32* (3), 291–297. https://doi.org/10.7334/psicothema2020.197
Frazier, B. (2006). The ethics of Rortian redescription. *Philosophy & Social Criticism 32* (4), 461–492. https://doi.org/10.1177/0191453706064020
Freud, S. (1955 [1917a]. Eine Schwierigkeit der Psychoanalyse. In A. Freud (Hrsg.), *Sigmund Freud: Gesammelte Werke. Chronologisch geordnet*. Zwölfter Band: Werke aus den Jahren 1917 – 1920 (11. Aufl., S. 3–12). London: Imago Publishing Co., Ltd.
Freud, S. (1955 [1917b]. *Sigmund Freud: Gesammelte Werke. Chronologisch geordnet* (11. Aufl., 7 Bände). London: Imago Publishing Co., Ltd. (Siebter Band: Werke aus den Jahren 1906–1909).
Fuchs-Heinritz, W. & König, A. (2014). *Pierre Bourdieu. Eine Einführung* (UTB, Bd. 2649, 3., überarb. Aufl.). Konstanz: UVK.
Gabriel, G. (1996). *Ästhetischer „Witz" und logischer „Scharfsinn". Zum Verhältnis von wissenschaftlicher und ästhetischer Weltauffassung*. Erlangen: Palm und Enke.
Gamm, G. (2011). Im Zwielicht des Versuchs. Experimentieren in Philosophie und Wissenschaften. In G. Gamm & J. Kertscher (Hrsg.), *Philosophie in Experimenten. Versuche explorativen Denkens* (S. 15–33). Bielefeld: transcript.
Gatzemeier, M. (2004). Energeia. In J. Mittelstraß (Hrsg.), *Enzyklopädie Philosophie und Wissenschaftstheorie* (1, A – G, unveränderte Sonderausgabe, S. 546). Stuttgart: J.B. Metzler.
Gethmann-Siefert, A. (2005). Drama oder Komödie? Hegels Konzeption des Komischen und des Humors als Paradigma der romantischen Kunstform. In A. Gethmann-Siefert, L. de Vos & B. Collenberg-Plotnikov (Hrsg.), *Die geschichtliche Bedeutung der Kunst und die Bestimmung der Künste* (Neuzeit & Gegenwart, S. 175–187). München: Wilhelm Fink.
Gibbons, M., Limoges, C., Nowotny, H., Schwartzmann, S., Scott, P. & Trow, M. (1994). *The New Production of Knowledge. The Dynamics of Science and Research in Contemporary Societies*. London: SAGE.
Gibson, J. J. (1982). *Wahrnehmung und Umwelt. Der ökologische Ansatz in der visuellen Wahrnehmung*. München: Urban & Schwarzenberg.
Gigerenzer, G. (2014). *Risiko. Wie man die richtigen Entscheidungen trifft*. München: btb.
Goldstein, E. B. & Cacciamani, L. (2023). *Wahrnehmungspsychologie. Der Grundkurs* (10. vollständig überarbeitete und aktualisierte Auflage). Berlin: Springer Gabler.

Grau, A. (2010). Von Witz, Humor und anderen komischen Dingen. *tv diskurs 14* (2), 18–23.

Hacke, J. (2021). Liberale Demokratie und Solidarität. Richard Rortys Hoffnung für den Westen. In V. Selk, C. Held & T. Schwuchow (Hrsg.), *Soziale Hoffnung, liberale Ironie. Zur Aktualität von Richard Rortys politischem Denken* (Staatsverständnisse, Bd. 145, S. 73–102). Baden-Baden: Nomos.

Hartmann, N. (1953). *Ästhetik*. Berlin: de Gruyter.

Hegel, G. W. F. (2003 [1823]. *Vorlesungen über die Philosophie der Kunst. Berlin 1823*. Hamburg: Meiner.

Heindl, A. (2023). Richard Rortys Kritik der Erkenntnistheorie: Demokratischer Anti- Repräsentationalismus. In M. Müller (Hrsg.), *Handbuch Richard Rorty* (1st ed. 2023, S. 581–596). Wiesbaden: Springer Fachmedien.

Hinz, L., Weber, A.-M., Koegst, L. & Kühne, O. (2024). A Neopragmatic Perspective on the Processual Nature of Landscape – Coastal Land Loss in Louisiana in the Context of Scientific Findings, Social Patterns of Interpretation, and Individual Experience. *Sustainability 16* (5), 1–26. https://doi.org/10.3390/su16052078

Hokema, D. (2015). Landscape is Everywhere. The Construction of Landscape by US-American Laypersons. *Geographische Zeitschrift 103* (3), 151–170.

Hradil, S. (2006). Soziale Milieus - eine praxisorientierte Forschungsperspektive, Bundeszentrale für politische Bildung. Aus Politik und Zeitgeschichte APuZ. https://www.bpb.de/shop/zeitschriften/ apuz/29429/soziale-milieus-eine-praxisorientierte-forschungsperspektive/. Zugegriffen: 20. February 2024.

Hradil, S. (2018). Milieu, soziales. In J. Kopp & A. Steinbach (Hrsg.), *Grundbegriffe der Soziologie* (12. Aufl., S. 319–322). Wiesbaden: Springer Fachmedien.

Hubig, C. (2006). *Die Kunst des Möglichen I. Technikphilosophie als Reflexion der Medialität*. Bielefeld: transcript.

Hubig, C. (2007). *Die Kunst des Möglichen II. Ethik der Technik als provisorische Moral* (Edition panta rei, Bd. 2). Bielefeld: transcript.

Hubig, C. (2015). *Die Kunst des Möglichen III. Macht der Technik* (Edition panta rei, Grundlinien einer dialektischen Philosophie der Technik 3). Bielefeld: transcript.

Jenal, C. (2019). *"Das ist kein Wald, Ihr Pappnasen!" – Zur sozialen Konstruktion von Wald. Perspektiven von Landschaftstheorie und Landschaftspraxis*. Wiesbaden: Springer VS.

Jurt, J. (2012). Bourdieus Kapital-Theorie. In M. M. Bergman, S. Hupka-Brunner, T. Meyer & R. Samuel (Hrsg.), *Bildung - Arbeit - Erwachsenwerden. Ein interdisziplinärer Blick auf die Transition im Jugend und jungen Erwachsenenalter* (S. 21–41). Wiesbaden: Springer.

Kahneman, D. (2014). *Schnelles Denken, langsames Denken*. München: Pantheon.

Kindt, T. (2017a). Humor. In U. Wirth (Hrsg.), *Komik. Ein interdisziplinäres Handbuch* (S. 7–11). Stuttgart: J.B. Metzler.

Kindt, T. (2017b). Komik. In U. Wirth (Hrsg.), *Komik. Ein interdisziplinäres Handbuch* (S. 2–6). Stuttgart: J.B. Metzler.

Knorr-Cetina, K. (2002). *Die Fabrikation von Erkenntnis. Zur Anthropologie von Wissenschaft*. Frankfurt (Main): Suhrkamp.

Koegst, L. (2022). Über drei Welten, Räume und Landschaften. Digital geführte Exkursionen an Hochschulen aus der Perspektive der drei Welten Theorie im Allgemeinen und der Theorie der drei Landschaften im Speziellen. *Berichte Geographie und Landeskunde 69* (3), 1–21. https://doi.org/10.25162/bgl-2022-0012

Kretz, D. (2023). Contingency, Irony, and Solidarity (1989). Übers.: Kontingenz, Ironie und Solidarität (1989). In M. Müller (Hrsg.), *Handbuch Richard Rorty* (1st ed. 2023, S. 201–224). Wiesbaden: Springer Fachmedien.

Kronenberg, B. (2014). *Die Zerbrechlichkeit des Wahren. Richard Rortys Neopragmatismus und Adornos Negative Dialektik* (Edition Moderne Postmoderne, 1st ed.). Bielefeld: transcript.

Küchler, J. & Wang, X. (2009). Vielfältig und vieldeutig. Natur und Landschaft im Chinesischen. In T. Kirchhoff & L. Trepl (Hrsg.), *Vieldeutige Natur. Landschaft, Wildnis und Ökosystem als kulturgeschichtliche Phänomene* (S. 201–220). Bielefeld: transcript.

Kuhn, T. S. (1970). *The structure of scientific revolutions* (2. Aufl., 2 Bände). Chicago: The University of Chicago Press. (Originalarbeit erschienen 1962).

Kühne, O. (2008). *Distinktion – Macht – Landschaft. Zur sozialen Definition von Landschaft.* Wiesbaden: VS Verlag für Sozialwissenschaften.

Kühne, O. (2017). *Zur Aktualität von Ralf Dahrendorf. Einführung in sein Werk* (Aktuelle und klassische Sozial- und Kulturwissenschaftler|innen). Wiesbaden: Springer VS.

Kühne, O. (2018a). *Landscape and Power in Geographical Space as a Social-Aesthetic Construct.* Dordrecht: Springer International Publishing.

Kühne, O. (2018b). Die Landschaften 1, 2 und 3 und ihr Wandel. Perspektiven für die Landschaftsforschung in der Geographie – 50 Jahre nach Kiel. *Berichte. Geographie und Landeskunde 92* (3–4), 217 – 231.

Kühne, O. (2020). Landscape Conflicts. A Theoretical Approach Based on the Three Worlds Theory of Karl Popper and the Conflict Theory of Ralf Dahrendorf, Illustrated by the Example of the Energy System Transformation in Germany. *Sustainability: Science, Practice and Policy 12* (17), 1–20. https://doi.org/10.3390/su12176772

Kühne, O. (2021). Der Dreifache Landschaftswandel als Herausforderung für Wissenschaft, Planung und Politik. In W. Strubelt, F. Dosch & G. Meinel (Hrsg.), *Die Gestalt des Raumes. Landschaften Deutschlands als Abbilder der Gesellschaft* (S. 92–94). Berlin: Wasmuth und Zohlen Verlag.

Kühne, O. (2024). *Redescribing Horizontal Geographies. A Neopragmatist Approach to Spatial Contingency, Complexity, and Relationships.* Cham: Springer International.

Kühne, O. & Jenal, C. (2020). The Threefold ´Landscape Dynamics – Basic Considerations, Conflicts and Potentials of Virtual Landscape Research. In D. Edler, C. Jenal & O. Kühne (Hrsg.), *Modern Approaches to the Visualization of Landscapes* (S. 389–402). Wiesbaden: Springer VS.

Kühne, O. & Leonardi, L. (2020). *Ralf Dahrendorf. Between Social Theory and Political Practice.* London: Palgrave Macmillan.

Kühne, O. & Berr, K. (2021). *Wissenschaft, Raum, Gesellschaft. Eine Einführung zur sozialen Erzeugung von Wissen.* Wiesbaden: Springer VS.

Kühne, O. & Koegst, L. (2022). Cartographic Representations of Coastal Land Loss in Louisiana. An Investigation Based on Deviant Cartographies. *KN – Journal of Cartography and Geographic Information 2022* (4), 1–12. https://doi.org/10.1007/s42489-022-00120-8

Kühne, O. & Koegst, L. (2023a). *Land Loss in Louisiana. A Neopragmatic Redescription.* Wiesbaden: Springer VS.

Kühne, O. & Koegst, L. (2023b). Neopragmatic Reflections on Coastal Land Loss and Climate Change in Louisiana in Light of Popper's Theory of Three Worlds. *Land 12* (2), 1–17. https://doi.org/10.3390/land12020348

Kühne, O., Berr, K., Jenal, C. & Schuster, K. (2021). *Liberty and Landscape. In Search of Life-chances with Ralf Dahrendorf*. Basingstoke: Palgrave Macmillan.

Kühne, O., Berr, K., Schuster, K. & Jenal, C. (2021). *Freiheit und Landschaft. Auf der Suche nach Lebenschancen mit Ralf Dahrendorf*. Wiesbaden: Springer.

Kühne, O., Berr, K. & Jenal, C. (2022). *Die geschlossene Gesellschaft und ihre Ligaturen. Eine Kritik am Beispiel ‚Landschaft'.* Wiesbaden: Springer VS.

Kühne, O., Koegst, L. & Berr, K. (2024). *Oilscapes of Louisiana – Neopragmatic Reflections on the Ambivalent Aesthetics of Landscape Constructions.* Wiesbaden: Springer Fachmedien.

Lakatos, I. (1974). Die Geschichte der Wissenschaften und ihre rationalen Rekonstruktionen. In I. Lakatos & A. Musgrave (Hrsg.), *Kritik und Erkenntnisfortschritt* (Abhandlungen des Internationalen Kolloquiums über die Philosophie der Wissenschaft, Bd. 4, S. 271–312). Braunschweig: Vieweg.

Lambert, C. E., Holley, J. R., McComas, K. A., Snider, N. P. & Tucker, G. K. (2021). Eroding Land and Erasing Place: A Qualitative Study of Place Attachment, Risk Perception, and Coastal Land Loss in Southern Louisiana. *Sustainability 13* (11). https://doi.org/10.3390/su13116269

Latour, B. & Woolgar, S. (2013 [1979]. *Laboratory Life: The Construction of Scientific Facts.* Princeton, New Jersey: Princeton University Press.

Leonardi, L. (2014). *Introduzione a Dahrendorf* (Maestri del Novecento, Bd. 20). Roma: Editori Laterza.

Lewin, K. (1927). Untersuchungen zur Handlungs- und Affektpsychologie. *Psychologische Forschung 9*(1), 1–85. https://doi.org/10.1007/BF02409755

Lichtenberg, G. C. (1973). *Sudelbücher. Schriften und Briefe* (Bd. 1). München: Carl Hanser Verlag.

Luhmann, N. (1995). *Die Kunst der Gesellschaft*. Frankfurt (Main): Suhrkamp.

Lützeler, H. (1999). *Rheinischer Humor. Nicht nur für Rheinländer*. Bonn: Bouvier.

Mackert, J. (2010). Opportunitätsstrukturen und Lebenschancen. *Berliner Journal für Soziologie 20*(3), 401–420. https://doi.org/10.1007/s11609-010-0135-7

Maischatz, K. (2010). Eine Einführung in das Sozialkapital-Konzept anhand der zentralen Vertreter. In A. Fischer (Hrsg.), *Die soziale Dimension von Nachhaltigkeit – Beziehungsgeflecht zwischen Nachhaltigkeit und Benachteiligtenförderung. Berufliche Bildung und zukünftige Entwicklung* (Leuphana-Schriften zur Berufs- und Wirtschaftspädagogik, Bd. 3, S. 31–54). Baltmannsweiler: Schneider-Verlag Hohengehren.

Malachowski, A. R. (2014). *Richard Rorty* (Philosophy now). Abingdon, Oxon: Routledge.

Mehrabian, A. (1987). *Räume des Alltags. Wie die Umwelt unser Verhalten bestimmt*. Frankfurt (Main): Campus.

Michaelis, R. (1997). Der lachende Anarchist, ZEIT online. https://www.zeit.de/1997/43/ fo.txt.19971017.xml/komplettansicht. Zugegriffen: 22. Juli 2020.

Müller, M. (2014). *Private Romantik, öffentlicher Pragmatismus? Richard Rortys transformative Neubeschreibung des Liberalismus* (Edition Moderne Postmoderne, 1st ed.). Bielefeld: transcript.

Musil, R. (1992 [1930]. *Der Mann ohne Eigenschaften* (Bd. 1, 2 Bände). Reinbek bei Hamburg: Rowohlt Taschenbuch Verlag.

Nickerson, R. S. (1998). Confirmation Bias: A Ubiquitous Phenomenon in Many Guises. *Review of General Psychology 2* (2), 175–220. https://doi.org/10.1037/1089-2680.2.2.175

Niedenzu, H.-J. (2001). Konflikttheorie: Ralf Dahrendorf. In J. Morel, E. Bauer, T. Maleghy, H.-J. Niedenzu, M. Preglau & H. Staubmann (Hrsg.), *Soziologische Theorie. Abriß ihrer Hauptvertreter* (7. Aufl., S. 171–189). München: R. Oldenbourg Verlag.

Nowotny, H., Scott, P. & Gibbons, M. (2001). *Re-Thinking Science. Knowledge and the Public in an Age of Uncertainty*. Cambridge: Polity Press.

Ogburn, W. F. (1922). *Social Change. With Respect To Culture And Original Nature*. New York: B. W. Huebsch.

Peng, L.-P. (2020). Understanding Human-Nature Connections Through Landscape Socialization. *International journal of environmental research and public health 17* (20), 7593. https://www.mdpi.com/862044.

Piaget, J. & Inhalder, B. (1975 [1947]. *Die Entwicklung des räumlichen Denkens beim Kinde*. Stuttgart: Klett-Cotta.

Popper, K. R. (1979). Three Worlds. Tanner Lecture, Michigan, April 7, 1978. *Michigan Quarterly Review 18* (1), 141–167.

Popper, K. R. (1996). *Alles Leben ist Problemlösen. Über Erkenntnis, Geschichte und Politik*. München: Piper.

Popper, K. R. & Eccles, J. C. (1977). *Das Ich und sein Gehirn*. München: Piper.

Poser, H. (2012). *Wissenschaftstheorie. Eine philosophische Einführung* (2., überarbeitete und erweiterte Auflage). Stuttgart: Philipp Reclam jun.

Preisendanz, W. (2019). Humor. In J. Ritter (Hrsg.), *Historisches Wörterbuch der Philosophie. Band 3: G-H* (Völlig neubearbeitete Ausgabe des »Wörterbuchs der philosophischen Begriffe« von Rudolf Eisler, Sonderausgabe, Sp. 1232–1234). Darmstadt: WBG.

Prominski, M. (2004). *Landschaft entwerfen. Zur Theorie aktueller Landschaftsarchitektur*. Berlin: Reimer.

Putnam, H. (1990a). *Die Bedeutung von „Bedeutung“* (2., durchgesehene Auflage). Frankfurt (Main): Klostermann.

Putnam, H. (1990b). *Vernunft, Wahrheit und Geschichte*. Frankfurt (Main): Suhrkamp.

Putnam, H. (1997). *Für eine Erneuerung der Philosophie* (Universal-Bibliothek, Bd. 9660). Stuttgart: Reclam.
Raab, E. (1981). Ästhetik und Neugier. In H.-J. Voss & H. Keller (Hrsg.), *Neugierforschung* (S. 263–308). Weinheim: Beltz.
Ramberg, B. T. (2011). Um seiner eigenen Generation Willen. Rorty über Zerstörung und Bildung. In A. Gröschner, M. Sandbothe & R. Rorty (Hrsg.), *Pragmatismus als Kulturpolitik. Beiträge zum Werk Richard Rortys* (Suhrkamp Taschenbuch Wissenschaft, Bd. 1981, 1. Aufl., Orig.-Ausg, S. 75–107). Berlin: Suhrkamp.
Rambow, R. (2000). *Experten-Laien-Kommunikation in der Architektur* (3. Aufl.). Münster: Waxmann.
Reese-Schäfer, W. (2016). *Richard Rorty zur Einführung*. Hamburg: Junius.
Rheinberger, H.-J. (2001). *Experimentalsysteme und epistemische Dinge. Eine Geschichte der Proteinsynthese im Reagenzglas*. Göttingen: Wallstein-Verlag.
Ritter, J. (1974 [1963]. *Subjektivität. Sechs Aufsätze*. Frankfurt (Main): Suhrkamp.
Rogers, C. R. (2005). *Die klientzentrierte Gesprächspsychotherapie. Client-Centered Therapy* (17. Aufl.). Fischer.
Rorty, R. (1980). *Philosophy and the mirror of nature* (1. paperback print). Princeton, N. J.: Princeton University Press.
Rorty, R. (1981). *Der Spiegel der Natur: Eine Kritik der Philosophie*. Frankfurt (Main): Suhrkamp.
Rorty, R. (1989). *Kontingenz, Ironie und Solidarität*. Frankfurt (Main): Suhrkamp.
Rorty, R. (1993). Feminism, Ideology, and Deconstruction: A Pragmatist View. *Hypatia 8* (2), 96–103. https://doi.org/10.1111/j.1527-2001.1993.tb00093.x
Rorty, R. (1997 [1989]. *Contingency, Irony, and Solidarity*. Cambridge: Cambridge University Press.
Rorty, R. (1999). *Philosophy and social hope* (Penguin books, 1. publ). London: Penguin Books.
Rorty, R. (2001). *Philosophie & die Zukunft. Essays* (2. Aufl.). Frankfurt (Main): Fischer.
Rorty, R. (2003). *Wahrheit und Fortschritt*. Frankfurt (Main): Suhrkamp.
Rorty, R. (2018 [1994]. *Hoffnung statt Erkenntnis. Eine Einführung in die pragmatische Philosophie* (3. Aufl.). Wien: Passagen-Verlag.
Rorty, R. (2019). *Philosophie als Kulturpolitik* (Suhrkamp Taschenbuch Wissenschaft, Bd. 2289, Erste Auflage). Berlin: Suhrkamp.
Rorty, R. (2023). *Pragmatismus als Antiautoritarismus*. Berlin: Suhrkamp.
Scherer, K. R. (2009). The dynamic architecture of emotion: Evidence for the component process model. *Cognition & Emotion 23* (7), 1307–1351. https://doi.org/10.1080/02699930902928969
Schlegel, F. (1963). *Philosophische Lehrjahre: 1796–1806. Nebst philosophischen Manuskripten aus den Jahren 1796–1828*. München, Paderborn, Wien: Schöningh.
Schlegel, F. (1967). *Kritische Friedrich-Schlegel-Ausgabe. Erste Abteilung. Kritische Neuausgabe* (Bd. 2). München, Paderborn, Wien, Zürich: Ferdinand Schöningh.
Schlegel, F. (1981). *Kritische Friedrich-Schlegel-Ausgabe. Fragmente zur Poesie und Literatur. Erster Teil*. München, Paderborn, Wien, Zürich: Ferdinand Schöningh.
Schlüter, D. (2019 [1971]. Akt/Potenz. In J. Ritter (Hrsg.), *Historisches Wörterbuch der Philosophie. Band 1: A-C* (S. 134–142). Darmstadt: WBG.
Schönhammer, R. (2013). *Einführung in die Wahrnehmungspsychologie. Sinne, Körper, Bewegung*. Stuttgart: UTB.
Schubert, H.-J., Joas, H. & Wenzel, H. (2010). *Pragmatismus zur Einführung. Kreativität, Handlung, Deduktion, Induktion, Abduktion, Chicago School, Sozialreform, symbolische Interaktion* (Zur Einführung, Bd. 382). Hamburg: Junius.
Schulze, G. (2000). *Die Erlebnisgesellschaft. Kultursoziologie der Gegenwart* (8. Aufl.). Frankfurt (Main): Campus.
Schulz von Thun, F. (2019). *Miteinander Reden 1-4: Störungen und Klärungen/Stile, Werte und Persönlichkeitsentwicklung/Das "Innere Team" und situationsgerechte Kommunikation/Fragen und Antworten*. Rowohlt Taschenbuch.

Schuster, K. (2021). Grundlagen der menschlichen Wahrnehmung und Stärken und Risiken visueller Kommunikation. In D. Bruns, B. Stemmer, D. Münderlein & S. Theile (Hrsg.), *Handbuch Methoden Visueller Kommunikation in der Räumlichen Planung* (S. 27–44). Wiesbaden: Springer Fachmedien.

Schuster, K. & Kirschbaum, M. (2011). Implikationen des soziologischen Lebensstilkonzepts auf die Architektur. Diskussion der Potenziale und Grenzen der Lebensstilforschung für die Architektur anhand einer empirischen Untersuchung. *Umweltpsychologie 15* (1), 90–111.

Schwemmer, O. (2004a). Akt und Potenz. In J. Mittelstraß (Hrsg.), *Enzyklopädie Philosophie und Wissenschaftstheorie* (1, A – G, unveränderte Sonderausgabe, S. 59–61). Stuttgart: J.B. Metzler.

Schwemmer, O. (2004b). Dynamis. In J. Mittelstraß (Hrsg.), *Enzyklopädie Philosophie und Wissenschaftstheorie* (1, A – G, unveränderte Sonderausgabe, S. 515–516). Stuttgart: J.B. Metzler.

Schwerz-Friesel, M., Marx, K. & Damisch, S. (2012). Persuasive Strategien der affektiven Verunsicherung im aktuellen Diskurs: Ironisieren, Kritisieren und Beleidigen in öffentlichen Streitgesprächen. In I. Pohl & H. Ehrhardt (Hrsg.), *Sprache und Emotion in öffentlicher Kommunikation* (S. 227–254). Frankfurt (Main): Peter Lang.

Spears, R. & Tausch, N. (2014). Vorurteile und Intergruppenbeziehungen. In K. Jonas, W. Stroebe & M. Hewstone (Hrsg.), *Sozialpsychologie* (6., vollständig überarbeitete Auflage, S. 507–564). Berlin: Springer.

Stamps, A. E. & Nasar, J. L. (1997). Design review and public preferences: Effects of geographical location, public consensus, sensation seeking, and architectural styles. *Journal of Environmental Psychology 17* (1), 11–32. https://doi.org/10.1006/jevp.1996.0036

Stotten, R. (2013). Kulturlandschaft gemeinsam verstehen – Praktische Beispiele der Landschaftssozialisation aus dem Schweizer Alpenraum. *Geographica Helvetica 68* (2), 117–127. https://doi.org/10.5194/gh-68-117-2013

Tautz, B. (2023). Klassiker der Ironie als Lebensform (Sokrates, Kierkegaard). In M. Müller (Hrsg.), *Handbuch Richard Rorty* (1st ed. 2023, S. 443–458). Wiesbaden: Springer Fachmedien.

Topper, K. (1995). Richard Rorty, Liberalism and the Politics of Redescription. *American Political Science Review 89* (4), 954–965. https://doi.org/10.2307/2082520

Treisman, A. (1993). The perception of features and objects. In A. D. Baddeley & L. Weiskrantz (Hrsg.), *Attention, selection, awareness and control. A tribute to Donald Broadbent* (S. 5–34). Oxford: Clarendon Press.

Vaihinger, H.-M. (1911). *Philosophie des als ob. System der theoretischen, praktischen und religiösen Fiktionen der Menschheit auf Grund eines idealistischen Positivismus.* Berlin: Reuther & Reichard (Mit einem Anhang über Kant und Nietzsche).

Vieweg, K. (2005). Heiterer Leichtsinn und fröhlicher Scharfsinn – Zu Hegels Verständnis von Komik und Humor als Formen ästhetisch-poetischer Skepsis. In A. Gethmann-Siefert, L. de Vos & B. Collenberg-Plotnikov (Hrsg.), *Die geschichtliche Bedeutung der Kunst und die Bestimmung der Künste* (Neuzeit & Gegenwart, S. 297–310). München: Wilhelm Fink.

Waldmann, M. R. (2018). Kategorisierung und Wissenserwerb. In J. Müsseler & M. Rieger (Hrsg.), *Allgemeine Psychologie* (S. 357–399). Berlin: Springer.

Weber, A.-M. (2024). The Social Construction of Land Loss in Louisiana. In K. Berr, L. Koegst & O. Kühne (Hrsg.), *Landscape Conflicts* (95-117). Wiesbaden: Springer VS.

Weber, B. (2023). Die liberale Ironikerin: Von der politischen Intellektuellen zur realistischen Träumerin. In M. Müller (Hrsg.), *Handbuch Richard Rorty* (1st ed. 2023, S. 883–899). Wiesbaden: Springer Fachmedien.

Weinrich, H. (2019). Ironie. In J. Ritter (Hrsg.), *Historisches Wörterbuch der Philosophie. Band 4: I-K* (Völlig neubearbeitete Ausgabe des »Wörterbuchs der philosophischen Begriffe« von Rudolf Eisler, Sonderausgabe, Spalten 577–582). Darmstadt: WBG.

Willer, S. (2017). Witz. In U. Wirth (Hrsg.), *Komik. Ein interdisziplinäres Handbuch* (S. 11–16). Stuttgart: J.B. Metzler.

Wirth, U. (2017). Ironie. In U. Wirth (Hrsg.), *Komik. Ein interdisziplinäres Handbuch* (S. 16–21). Stuttgart: J.B. Metzler.

Wojtkiewicz, W. & Heiland, S. (2012). Landschaftsverständnisse in der Landschaftsplanung. Eine semantische Analyse der Verwendung des Wortes „Landschaft“ in kommunalen Landschaftsplänen. *Raumforschung und Raumordnung – Spatial Research and Planning 70* (2), 133–145. https://doi.org/10.1007/s13147-011-0138-7

Zhang, K., Zhao, J. & Bruns, D. (2013). Landschaftsbegriffe in China. In D. Bruns & O. Kühne (Hrsg.), *Landschaften: Theorie, Praxis und internationale Bezüge. Impulse zum Landschaftsbegriff mit seinen ästhetischen, ökonomischen, sozialen und philosophischen Bezügen mit dem Ziel, die Verbindung von Theorie und Planungspraxis zu stärken* (S. 133–150). Schwerin: Oceano Verlag.

Chapter 3
Concept: Inverse Landscapes and Their Representations—From Contingency to Irony

In the sense of 'ironistic theorizing', is the concept of an 'inverse landscape'—which, as shown, expresses the awareness of the contingency of landscape (on all three levels). This concept was developed in the early 2010s (Kühne 2013b, 2013c) and was given further focus in the context of the presentation of oilscapes as well as the theoretical considerations on the formulation of a neopragmatist research program for dealing with 'horizontal geographies' (i.e., spatial syntheses as undertaken by regional geography) (Kühne et al. 2023a, 2024; Kühne 2024). This last stage of expansion is now the starting point for further (theoretical) engagement with the question of operationalizing the concept of contingency for spatial science research.

3.1 Inverse Landscapes—Basic Considerations

The concept of inverse landscapes took its starting point in the context of dealing with the complex relationships between landscape and power (Kühne 2012, 2013b, 2013c). The expansion at the beginning of the 2020s involved the explicit integration of the (not solely power-related) contingencies of Landscape 1, Landscape 2, and Landscape 3 (Kühne et al. 2023a, 2024; Kühne 2024). The starting point for the reflections on the concept of the inverse landscape can be found in the considerations of the German geographer Gerhard Hard (2008, S. 268) on manifestations and non-manifestations, especially of the societally oriented Landscape 1, in which the theme of contingency is already present: In "every reasonably good, i.e., also: methodologically reflected (social) geographical excursion, the question is (today hopefully by default) not only what of society, economy, ecology and history one can see or access in space, in the landscape, in the terrain, but also what one does not see for what reason (although precisely this might be far more important for the given topic)—and what one might only think one sees because one has projected one's possibly

O. Kühne et al., *Neopragmatism - Inverse Landscape - (Carto)graphic Representation*, Neopragmatic Horizontal Geographies 2, https://doi.org/10.1007/978-3-031-89402-2_3

even false prior knowledge onto something that might mean something completely different".

However, the conceptual roots of thinking about contingency go back even further: Korf (2023) points out in his reflections on "Geography in the Subjunctive" that Ferdinand von Richthofen had already posed the question of "what if" in his inaugural lecture at the University of Leipzig in 1883. However, these considerations conveyed and still relate to the question regarding the vocabulary of geography as a discipline. In dealing with contingent and inverse landscapes, we go beyond this. We also pursue – with the neopragmatist approach as a whole—the goal of offering scientific geography (and beyond) a redescriptive vocabulary. However, we also relate our considerations to the construction of spaces in general and landscapes in particular. Furthermore, we conceptually include the presentation of inverse landscapes (also spaces) in our considerations. In addition, we draw on a decidedly elaborated philosophical metatheoretical framework, which is particularly oriented towards the enhancement of contingency.

In the early conception of inverse landscapes, the aim was to point out how power-bound processes were reflected in the social construction of landscape, in their individual experience as well as in the physical world. The manifest landscapes (on all three levels) that emerged as a result of the use of the 'opportunity to assert one's own will' (Weber 1976 [1922] were contrasted with inverse landscapes in order to show which landscapes could not be develop because of 'insufficient power' (Paris 2005).

This idea was generalized in the second phase of exploring inverse landscapes, so that the issue of power no longer has a constitutive function.

In this respect, inverse landscapes (on all three levels) can be understood as a concretization of contingent landscapes; they thus become clear as a possibility, as they visualize something from a diffuse space. Contingent landscapes can be understood as those landscapes that cannot be found in this form, but whose manifestation would not be impossible. Inverse landscapes are a concrete example of contingent landscapes. While contingent landscapes remain in a rather diffuse space of possibilities, inverse landscapes can be condensed into a concrete idea of a contingent state and thus concretized as one possibility (among many others)—and represented (carto)graphically.

Contingent, as well as inverse, landscapes connect to Richard Rorty's understandings of 'therapeutic redescriptions' and 'historical reconstructions'. Therapeutic redescriptions are "a method to illustrate the idea of de-ontologizing reality and to strengthen the readers' insight into their constructive ability in the discursive formation of ontic reality" (Merlingen 1999, p. 98). By 'historical reconstructions' is meant what "non-reeducated, deceased thinkers would have said to their contemporaries" (Rorty 2003, p. 362). Both concepts pursue the goal of visualizing contingency in the form of possible events that did not occur in this form. Another "philosopher of contingency" (Dries 2009, p. 89)—namely Günther Anders—was also on the trail of this idea, insofar as he wrote in the 1940s about so-called 'what-if films' ('IF pictures'): "After all, each historical situation contains various possibilities and plans, each of which has a certain degree of probability, each of which could become reality, before it is excluded by the one which turns out to be victorious" (Dries 2009, p. 89). Contingent and inverse landscapes also pursue the goal of visualizing contingency. The edge of contingency is determined by fantasy, because "reason cannot leave the final circle drawn by fantasy. In this sense, imagination takes precedence over reason" (Rorty 2019, p. 204).

3.2 The Expansion of 'Landscape' Understanding by Means of Inverse Landscapes

Based on these explanations, five landscape terms can be defined (Fig. 3.1): Landscape 3 as those interpretations, evaluations, and categorizations of landscape that are commonly shared socially. Contingent Landscape 3 includes those interpretations, evaluations, and categorizations of landscape that are partly socially shared, partly not accepted, but partly also taken into consideration. Between these two Landscapes 3, the tipping point or—with recourse to Kühne and Fischer (2022)—Pillepallerand[1] can be described. Pillepalle can be used to describe something whose meaning is interpreted differently. While a specific 'something' has great significance for one person, this something does not have the same significance for another person, who in turn (sometimes with a distinctive intention) describes it as 'Pillepalle'. The fact that things that are described as 'trivial' are not things of existential importance becomes clear from the possibility of extremely different evaluations. After people (as farmers, for example) were increasingly no longer existentially dependent on landscape and an aestheticized construction became widespread, landscape relegated into an area within the pattern of interpretation of 'Pillepalle'. Correspondingly, there were extensive disputes—especially in a Mode C context—about questions of the characterization of certain Spaces 1 as landscape, the integration of certain objects into an understanding of landscape, and not least about understandings of landscape itself. What the disputants had in common was an existential distance from landscape (Kühne und Fischer 2022; on the term Pillepalle: Bedeutung Online 2019; Forberg 2020; Heine 2019; Ruhrgebietssprache 2013). The Kippsaum[2] or Pillepallerand is an area in which inverse landscapes (as concretized contingent landscapes) are increasingly transformed into Landscape 3. In addition, a boundary is devised between contingent landscape and 'Not-Landscape 3', i.e., 'something' that must under no circumstances (without having to fear the loss of social recognition) be referred to as landscape.

[1] Pillepallerand, Pillepalle: German colloquialisms, common in the Ruhr region. In general, Pillepalle conveys a trifling matter/insignificance (also, something easily obtained), Pillepallerand conveys being on the fringe or margins of "Pillepalle" – in the context of usage here, as neither significant nor insignificant depending (contingent) upon the specific values and worldviews of the individual (or entity) speaking and, equally, those of the listener.

[2] Kippsaum: a German colloquialism, common in the Ruhr region, generally conveying "a precipice", "tilting edge", or "tipping point".

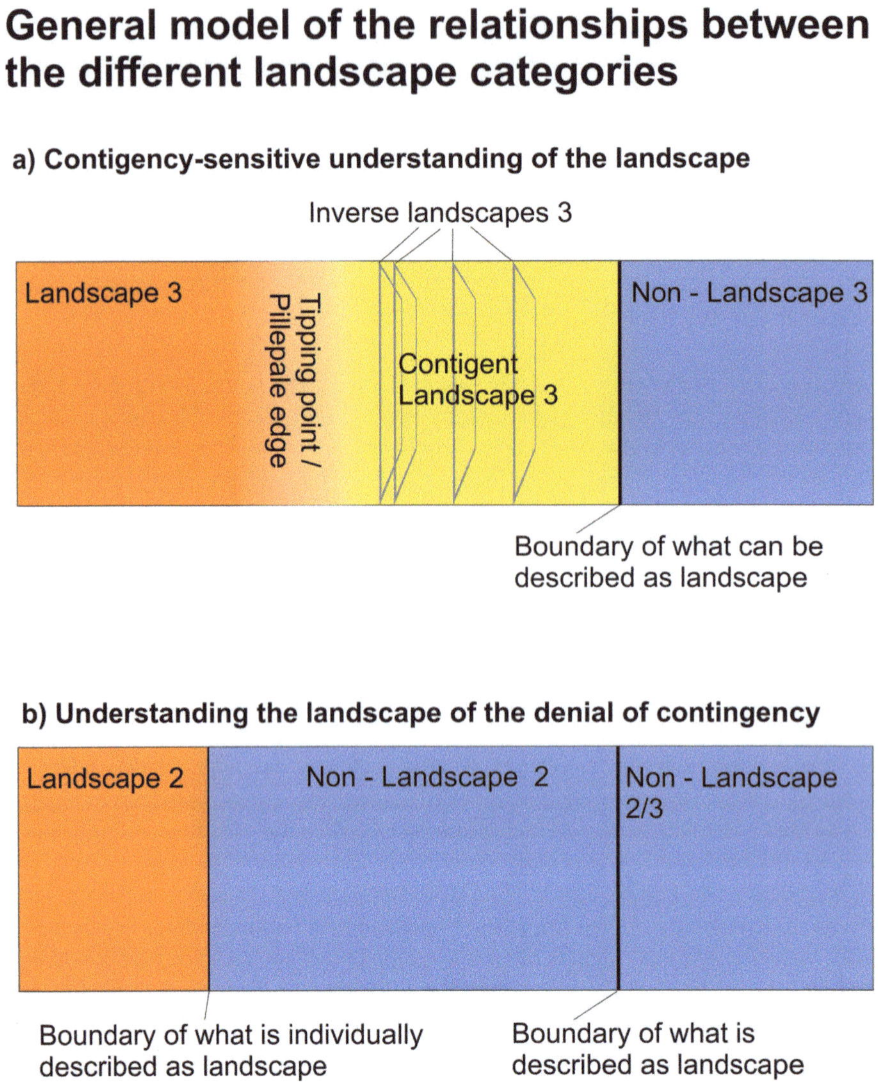

Fig. 3.1 The contingency-aware understanding of landscape and the landscape understanding of contingency denial (own illustration)

In addition to the contingency-sensitive understanding of landscape, there are also understandings of landscape (often in Mode C) that strive to destroy contingency (see Fig. 3.1). Here, a clear boundary is drawn between Landscape 2 and 'Not-Landscape 2'. Configurations that are presented as 'contingent landscapes' in Fig. 3.1 are desinated as Not-Landscape 2. Landscape 2 is subjected to binary coding here. Such a categorical rejection of contingency exhibits a high degree of stability. If contingency is individually admitted, the individual understanding of landscape can be expanded by integrating an inverse of Landscape 2 as a result of one's own experience of a Space 1 as (inverse) Landscape 1 (Fig. 3.2).

3.3 The Temporality of Landscapes—Contingent Landscapes and Inverse Landscapes

Based on the understanding of the contingency of landscape on all three levels, the following differentiation can be made, which focuses on the past, present, and future (based on Kühne et al. 2023a, 2024; Kühne 2024): Inverse Landscape 1 comprises those possible objects and object configurations that were not physically manifest, but which could certainly have manifested in the respective spatiotemporal framework. This non-manifestation may not have occurred for various reasons resulting from different degrees of the hybridity involving nature and culture, for example, if a certain land use were to prove to be no longer profitable as a result of changing ecological conditions or if economic conditions were to change (a product can be manufactured more cost- effectively elsewhere). Even if power no longer plays a constitutive role in the second phase of concept development, it is still significant; albeit less in the Weberian sense of power as an opportunity and more in the Rortyian sense of power as a pan-relational concept. In this way, the pan-relational power relationships can shift. Accordingly, power relationships can shift in such a way that an area of land must be subject to regulations of nature conservation, for example, with the result that certain agricultural practices are no longer permitted. Although power relationships are omnipresent, they are certainly reversible. Power relationships are also reversible because power is not only tolerated, but the experience of power can be contemplated. Critical engagement with power can be used creatively for self-empowerment (Allen 2008; Dewey 1958; Rorty 1991). This Landscape 2 oriented engagement with the connection between power and landscape can, not only, expand the self-empowerment of World 2, but also produce new contingent and inverse landscapes (Landscape 3). If these can assert themselves in the relational power structure, they can also be concretized as a Landscape 1. Thus, an inverse Landscape 1 becomes manifest as a Landscape 1. The inversion of Landscape 3 refers to its creation from different vocabularies of Mode B and especially Mode C, which are created from contingent notions of landscape. This applies to a lesser extent to the Mode A, whose contribution to the creation of inverse landscapes is primarily limited to sentimentally remembered, long-lost landscapes. In principle, the boundaries of inverse landscapes follow the boundaries of contingent landscapes and are—as shown above—bound to imagination. In relation to inverse landscapes, this means that where imaginative constructions of landscape can no longer be materialized, the edge of contingency is reached. As a result of the large and growing contingency, the number of inverse landscapes is growing.

As indicated, landscape in general and inverse landscape (on all three levels) are conditioned temporally in three ways:

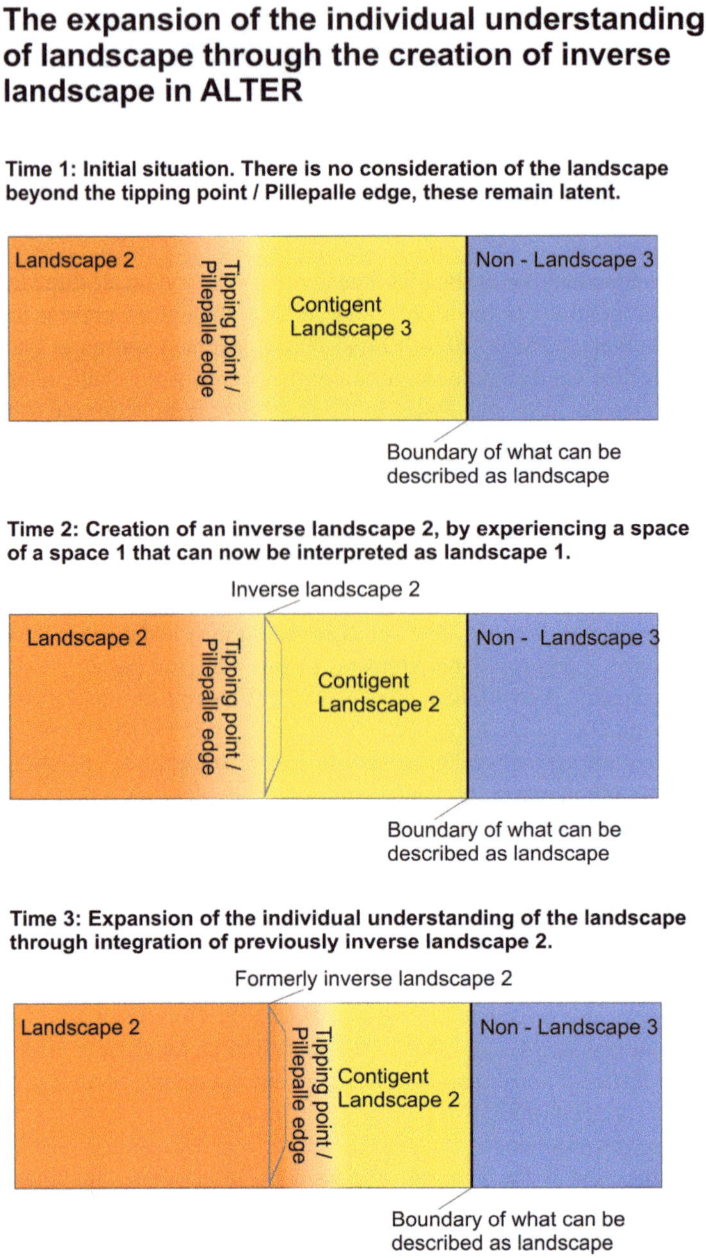

Fig. 3.2 The extension of 'Landscape 2' through the integration of inverse landscape (own illustration)

1. Landscapes are an expression of the social conditions in which they were created. However, many of these conditions are no longer to be found in this form today, especially tangible manifestations that exhibit great physical persistence, such as bunkers, old industrial plants, etc. (e.g., Fehn 1976; Schenk 2011), illustrating historical landscape contingency. Persistence is not limited to objects and object constellations of Landscape 1. It can also be found in the content of Landscape 3. For example, ideas of a 'romantic landscape' still exhibit considerable persistence today. These hardly correspond to the current 'requirements' (e.g., of an economic nature) of landscape (Landscape 1) in today's world (see, for example: Kirchhoff 2011, 2017; Kirchhoff und Trepl 2009; Kühne 2013a; Kühne und Berr 2024), today they form a contingence to inverse landscape (depending on the degree of manifestation), against which today's landscape development is (aesthetically) measured.
2. Landscapes can be understood as primary and secondary consequences of social structures and processes. The treatment of landscape appeared appropriate in the respective (!) past. Landscapes, whether non-inverse or inverse, can always be understood as a temporally contextualized specific engagement with space in general and landscape in particular.
3. Landscapes 3—and even more so inverse Landscapes 3—have the potential to point to the future. This becomes particularly clear in the context of spatial planning. Spatial planning deals with contingent states and outlines inverse states, among which it selects and normatively charges one. After all, planning not only involves the currently existing, but aims to meet future (and present) challenges (Healey 2009). However, the construction of inverse future landscapes does not have to exhibit this degree of normativity and embodiment, but can also be more speculative, dystopian, and utopian (whereby certain worldviews often become clear here).

Inverse landscapes are ultimately the result of deviated constructions of landscape, of space in general, which are carried out in World 2—with recourse to the contents of World 3 in projection onto World 1 (for more details on cartography: Edler und Kühne 2022b, 2023; Kühne 2021b; Kühne und Koegst 2022). Inverse landscapes are based on assumptions, on this side of the impossible and beyond what has occurred. They can thereby have a functional effect if they expand contingency. They can have a meta-functional effect if they expand ideas about landscape, the willingness to reflect on landscape and its contingency. But they can also be afunctional if the construction of inverse landscapes has no social resonance. However, inverse landscapes can also be dysfunctional if—despite their constitutive connection to contingent landscapes—they attempt to limit the contingency of landscape concepts.

To summarize succinctly, the temporal boundedness of the concept of inverse landscapes can be encapsulated as follows: Inverse landscapes can be understood as landscapes that have already existed, which cannot be inferred from current functions and structures of society. They can also be understood as landscapes that have not occurred yet, i.e., landscapes that have not found a manifestation. This also applies to future ideas of landscape that are currently still foreseeable; they too are inverse. As a result of educational expansion, an increasing production of knowledge involving landscape (especially in Mode C), and social differentiation (with an increase in perspectives), the contingency space in which inverse landscapes can be developed is growing. This can be interpreted—based on the optimistic understanding of power, as is inherent in neopragmatism—as an expression of an active and empowering interpretation of power. By drawing on the attitude of irony, the competition of inverse landscapes can be productively shaped: Here, the distancing of the inauthentic from the object of construction is always maintained.

3.4 Inverse Landscapes and Their (Carto)graphic Representation

Inverse landscapes can be understood as embodied projections of spatiotemporal contingency. They arise from an awareness of the contingency of self, language, and society. We pursue a broad understanding of language that not only considers spoken and written words as language, but also other signs that are expressed linguistically, including cartographic symbology, which are also referred to as signature language (Arnberger 1993). This awareness of the contingency of self, (signature) language, and society finds its expression in irony. This has consequences for the (carto)graphic realization of inverse landscapes. This is based on a design that distances itself from a representational understanding of the 'recognizing' subject and the 'recognized' object. Instead, the clarification of contingency takes on a prominent role, this in connection with the pragmatist principle of not being limited to the creation of pictorial, multimedia, and multimodal presentations (Edler et al. 2019; Endreß 2023; Krygier 1994) to focus on what contributes to the understanding of what is to be expressed. This means, for example, that information from topographical and cartographic bases that also have no relevant (i.e., contextualizing) connection to the central object or topic of the investigation should be avoided. To make it clear that the context of the investigation, as well as the cartographic forms of representation, are contingent, it is advisable to approach the design with an attitude of irony and to express this in what is produced. This in turn can be achieved through unconventionality. This relates to the topic, the type of data preparation, particularly the use of multidimensional signature language or other visual language. One approach, for example, is to allow photos (which often suggest 'objectivity') to take a back seat to the use of collages or comic-like representations to illustrate contingency. This example results in an unconventional interpretation of generalization in the pictorial representation of spatial information, as the reduction of the graphic objects (as is usual in cartographic generalization) transforms the ostensible subjectivity into a more objective representation. From a neopragmatist perspective, it also becomes relevant to justify the choice of topics, data, and forms of presentation to readers to ensure transparency regarding one's own considerations—and to provide an opportunity to engage meta-functionally with the topic and its presentation (Kühne et al. 2023a, 2024; Kühne 2023).

A common practice of processing spatial information is that of cartography (among many: Dickmann 2018; Edney 2019; Field 2018; Freitag 2008; Monmonier 1993; Ricker 2020; Schiewe 2022).

Cartography can be understood as a praxis at the interface of natural sciences, engineering, and the humanities, which emerged endeavoring to reduce the complexity of everyday life. Cartographic abstraction emerged based on 'common sense' and everyday practical needs (e.g., graphic fixation of borders and support for fiscal regulations) and thus showed little affinity for ironic thinking.

Moreover, the modernist tendency towards standardization and unification meant an attempt to destroy contingency (Bagrow 2017; Edler und Kühne 2022a; Kent 2005, 2013). 'Critical cartography' interpreted this urge to destroy contingency—here in the neopragmatic, contingency-oriented terminology introduced above—as an expression of the attempt to use power to prevent people from developing contingent perspectives on the world (for example in: Counter Cartographies Collective et al. 2012; Crampton und Krygier 2005; Harley 1989; Kim 2015; Wood und Krygier 2009). This is an endeavor that was itself associated with (possibly unintended) diminutions of contingent perspectives. For the rejection of 'traditional positivist cartography', most succinctly in Wood (2003), does not mean weaving a 'time-honored vocabulary' into a vocabulary that is more suitable for current conditions, but rather the attempt to simply displace the 'time-honored vocabulary' from the world, thereby misjudging its social utility (cf. Edler und Kühne 2022b; Kühne 2021a). There also seems to be little to no ironic approaches in critical cartography, unlike the neopragmatic scientific perspective represented here.

A neopragmatic cartography draws on the romantic and in turn the postmodern concept of irony as an attitude. Accordingly, it can be understood—as well as the cartographic representations arising from this understanding—as redescriptions of the world. From these follows (in manifestation of the representations in Sect. 2.1):

1. At the time of their creation (!), cartographic representations must be more suitable for interpreting the 'world' than cartographic representations created to date.
2. Cartographic representations must be open to providing content, forms of representation or perspectives that can be taken up and further developed in future cartographic representations (Rorty 1997 [1989]; Topper 1995).
3. Cartographic representations must interpret the world contingently by describing contingencies, on the one hand, but also by producing contingent interpretations, on the other.

However, cartography representations are not only to be assigned to the private sphere (i.e., the sphere of irony), but also have an effect in the public sphere. In this respect, they (or their authorship) have the task of 'reducing cruelty', as Richard Rorty echoes Judith Shklar. Thus, according to Rorty (1997 [1989]), the most effective way to inflict lasting pain on people is to humiliate them. Humiliation is particularly effective when what was, and is, particularly important to people is made to appear futile, outdated, or powerless. The aim of an inclusive development of society is therefore to avoid such humiliations (Rorty 1997 [1989]) and "to make the evils of cruelty the basic standard of its political practices and regulations. The only exception to this rule is the prevention of even greater cruelty" (Shklar 2020, p. 46).

The effort to reduce cruelty and humiliation concerns not only the content (which can be suitable for strengthening resentment, for example), but also the type of cartographic representations (such as a disrespectful symbology). However, a neopragmatic approach that is sensitive to contingencies can be made appreciably clear using ironic representations. After all, irony is particularly suitable for highlighting the contingency, provisional nature, and revisability of redescriptions (Edler und Kühne 2022b; Kühne et al. 2021, 2022; Kühne und Jenal 2020). New techniques for generating cartographic representations, such as augmented and virtual realities, offer significant potential here. Certain aspects of ironic cartography can be derived from the above (see in particular Kühne et al. 2023b; Kühne et al. 2024; Kühne 2024:

a. Self-ironic cartography in which a cartographer distances himself or herself.
b. Irony to the conventions of cartography, in that exaggerations and understatements are made in deviation from cartographic forms of representation.
c. Irony as the subject of cartographic representation, in which we find exaggerations and understatements in relation to what is depicted, such as caricature maps.
d. Meta-ironic cartographies that find themselves at a reflexive distance from the conventions of knowledge production (such as the positivist paradigm).
e. Irony in relation to the choice of 'mappers', i.e., in relation to the fact that certain conventions of cartography are unknown. Here the transition to cynicism is fluid.

Point e (above) makes it especially clear that ironic cartography certainly has areas of intersection with other, particularly comic, approaches to the world. The central difference, also to humorous cartography, lies in the fact that ironic cartography is not limited to distancing itself by means of exaggeration or understatement, but—in our understanding—is linked to a component of generating contingency, i.e., an increase in perspectivity and possibilities of understanding oneself, language, society, and other, interwoven worlds. Since ironic cartography constitutively seeks to expand contingency, it distinguishes itself from attempts to create cartographic representations that seek to demonstrate the 'essence' of a described space, but also from representations that seek to reduce contingency by mocking it. In this respect, the shift towards cynicism, as the dark side of irony, requires special sensitivity. The hybridization of the above-mentioned types of ironic cartography can offer a way out of this, for example, if a cartographic representation that can also be read as cynical (Point e, above)—without an unusually pronounced sensitivity—is interwoven with a special self- irony (Point a, above).

References

Allen, J. (2008). Pragmatism and power, or the power to make a difference in a radically contingent world. *Geoforum 39* (4), 1613–1624. https://www.sciencedirect.com/science/article/pii/S00167 18507000978.

Arnberger, E. (1993). *Thematische Kartographie* (3. Aufl.). Braunschweig: Westermann.

Bagrow, L. (2017). *History of Cartography* (2. Aufl.). London: Taylor & Francis.

Bedeutung Online. (2019). Was bedeutet Pillepalle (Pille palle)? Bedeutung, Definition, Wortherkunft - Bedeutung Online. https://www.bedeutungonline.de/was-bedeutet-pillepalle-pille-palle-bedeutung-definition-wortherkunft/. Zugegriffen: 30. Juli 2022.

Counter Cartographies Collective, Dalton, C. & Mason-Deese, L. (2012). Counter (Mapping) Actions: Mapping as Militant Research. *ACME: An International Journal for Critical Geographies 11* (3), 439–466.

Crampton, J. & Krygier, J. (2005). An Introduction to Critical Cartography. *ACME: An International Journal for Critical Geographies 4* (1), 11–33.

Dewey, J. (1958). *Art as experience* (Capricorn books, Bd. 1, 12. impr). New York, NY: Putnam.

Dickmann, F. (2018). *Kartographie* (Das Geographische Seminar). Braunschweig: Westermann.

Dries, C. (2009). *Günther Anders*. Paderborn: W. Fink.

Edler, D. & Kühne, O. (2022a). Aesthetics and Cartography: Post-Critical Reflections on Deviance in and of Representations. *ISPRS – International Journal of Geo-Information 11* (10). https://doi.org/10.3390/ijgi11100526

Edler, D. & Kühne, O. (2022b). Deviant Cartographies: A Contribution to Post-critical Cartography. *KN – Journal of Cartography and Geographic Information 72*, 103–116. https://doi.org/10.1007/s42489-022-00110- w

Edler, D. & Kühne, O. (2023). Post-Critical Cartography and Deviance: Overcoming Dysfunctional Aspects of Critical Cartography?! *Abstracts of the International Cartographic Association (ICA) 6* (58), 1. https://doi.org/10.5194/ica-abs-6-58-2023

Edler, D., Kühne, O., Keil, J. & Dickmann, F. (2019). Audiovisual Cartography: Established and New Multimedia Approaches to Represent Soundscapes. *KN – Journal of Cartography and Geographic Information 69*, 5–17. https://doi.org/10.1007/s42489-019-00004-4

Edney, M. H. (2019). *Cartography. The ideal and its history*. Chicago: The University of Chicago Press.

Endreß, S. (2023). A Novel Approach to Visualizing Olfactory Landscapes. *KN – Journal of Cartography and Geographic Information*, 1–12. https://doi.org/10.1007/s42489-023-00144-8

Fehn, K. (1976). Historische Geographie. Eigenständige Wissenschaft und Teilwissenschaft der Geographie. *Mitteilungen der Geographischen Gesellschaft München* (61), 35–51.

Field, K. (2018). *Cartography. A compendium of design thinking for mapmakers* (1. Aufl.). Redlands, California: ESRI Press.

Forberg, F. (2020). Das Pillepalle, Deutsche Welle (www.dw.com). https://www.dw.com/de/das-pillepalle/a-48252032. Zugegriffen: 30. Juli 2022.

Freitag, U. (2008). Von der Physiographik zur kartographischen Kommunikation – 100 Jahre wissenschaftliche Kartographie. *KN – Journal of Cartography and Geographic Information 58* (2), 59–67. https://doi.org/10.1007/BF03543975

Hard, G. (2008). Der Spatial Turn, von der Geographie her beobachtet. In J. Döring & T. Thielmann (Hrsg.), *Spatial Turn. Das Raumparadigma in den Kultur- und Sozialwissenschaften* (S. 263–316). Bielefeld: transcript.

Harley, J. B. (1989). Deconstructing The Map. *Cartographica: The International Journal for Geographic Information and Geovisualization 26* (2), 1–20. https://doi.org/10.3138/E635-7827-1757-9T53

Healey, P. (2009). The Pragmatic Tradition in Planning Thought. *American Journal of Agricultural Economics 28* (3), 277–292. https://doi.org/10.1177/0739456X08325175

Heine, M. (2019, 10. Juni). Ein Mann, ein Wort: Merkels Pillepalle – vom Szenewort ins Politdeutsch. *WELT*. https://www.welt.de/kultur/article195023759/Ein-Mann-ein-Wort-Merkels-Pillepalle-vom- Szenewort-ins-Politdeutsch.html. Zugegriffen: 12. September 2022.

Kent, A. J. (2005). Aesthetics: A Lost Cause in Cartographic Theory? *The Cartographic Journal 42* (2), 182–188. https://doi.org/10.1179/000870405X61487

Kent, A. J. (2013). From a dry statement of facts to a thing of beauty. Understanding aesthetics in the mapping and counter-mapping of place. *Cartographic Perspectives 73,* 39–60. https://repository.canterbury.ac.uk/item/87062/from-a-dry-statement-of-facts-to-a-thing-of-beauty-understanding-aesthetics-in-the-mapping-and-counter-mapping-of-place.

Kim, A. M. (2015). Critical cartography 2.0: From "participatory mapping" to authored visualizations of power and people. *Landscape and Urban Planning 142,* 215–225. https://doi.org/10.1016/j.landurbplan.2015.07.012

Kirchhoff, T. (2011). ‚Natur' als kulturelles Konzept. *Zeitschrift für Kulturphilosophie 5* (1), 69–96.

Kirchhoff, T. (2017). Landschaft. In T. Kirchhoff, N. C. Karafyllis, D. Evers, B. Falkenburg, M. Gerhard, G. Hartung et al. (Hrsg.), *Naturphilosophie. Ein Lehr- und Studienbuch* (S. 152–158). Tübingen: Mohr Siebeck; UTB.

Kirchhoff, T. & Trepl, L. (2009). Landschaft, Wildnis, Ökosystem: zur kulturbedingten Vieldeutigkeit ästhetischer, moralischer und theoretischer Naturauffassungen. Einleitender Überblick. In T. Kirchhoff & L. Trepl (Hrsg.), *Vieldeutige Natur. Landschaft, Wildnis und Ökosystem als kulturgeschichtliche Phänomene* (S. 13–68). Bielefeld: transcript.

Korf, B. (2023). German Theory als Geographie im Konjunktiv, oder: „Was nie geschrieben wurde, lesen". *Geographica Helvetica 78* (3), 325–336. https://doi.org/10.5194/gh-78-325-2023

Krygier, J. B. (1994). Sound and Geographic Visualization. In A. M. MacEachren & D. R. F. Taylor (Hrsg.), *Visualization in Modern Cartography* (Bd. 2, S. 149–166). Oxford: Elsevier.

Kühne, O. (2012). *Stadt – Landschaft – Hybridität. Ästhetische Bezüge im postmodernen Los Angeles mit seinen modernen Persistenzen.* Wiesbaden: Springer VS.

Kühne, O. (2013a). Alte Träume, neue Fluchten? Romantische Rezepte für die Gegenwart? Impulse zur Wiederentdeckung der Romantik in der Landschafts- und Siedlungsgestaltung zwischen Main und Rhein: ein Ausblick. In Verwaltung der staatlichen Schlösser und Gärten Hessen (Hrsg.), *RheinMainRomantik. Gartenkunst* (S. 291–300). Regensburg: Schnell & Steiner.

Kühne, O. (2013b). Landschaft zwischen Objekthaftigkeit und Konstruktion – Überlegungen zur inversen Landschaft. In D. Bruns & O. Kühne (Hrsg.), *Landschaften: Theorie, Praxis und internationale Bezüge. Impulse zum Landschaftsbegriff mit seinen ästhetischen, ökonomischen, sozialen und philosophischen Bezügen mit dem Ziel, die Verbindung von Theorie und Planungspraxis zu stärken* (S. 181–193). Schwerin: Oceano Verlag.

Kühne, O. (2013c). Landschaftsästhetik und regenerative Energien – Grundüberlegungen zu De- und Re-Sensualisierungen und inversen Landschaften. In L. Gailing & M. Leibenath (Hrsg.), *Neue Energielandschaften – Neue Perspektiven der Landschaftsforschung* (S. 101–120). Wiesbaden: Springer VS.

Kühne, O. (2021a). Contours of a 'Post-Critical' Cartography – A Contribution to the Dissemination of Sociological Cartographic Research. *KN – Journal of Cartography and Geographic Information 71,* 133–141. https://doi.org/10.1007/s42489-021-00080-5

Kühne, O. (2021b). Potentials of the Three Spaces Theory for Understandings of Cartography, Virtual Realities, and Augmented Spaces. *KN – Journal of Cartography and Geographic Information 71* (4), 297–305. https://doi.org/10.1007/s42489-021-00089-w

Kühne, O. (2023). Foodscapes – a Neopragmatic Redescription. *Berichte. Geographie und Landeskunde 96* (1), 5–25. https://doi.org/10.25162/bgl-2022-0016

Kühne, O. (2024). *Redescribing Horizontal Geographies. A Neopragmatist Approach to Spatial Contingency, Complexity, and Relationships.* Cham: Springer International.

Kühne, O. & Jenal, C. (2020). Baton Rouge (Louisiana): On the Importance of Thematic Cartography for 'Neopragmatic Horizontal Geography'. *KN – Journal of Cartography and Geographic Information 71* (1), 23-31. https://doi.org/10.1007/s42489-020-00054-z

Kühne, O. & Fischer, J. D. (2022). Pillepallescapes. Journeys to the edges of socially shared landscape constructions. *Human Geographies – Journal of Studies and Research in Human Geographies 16* (2), 157–174. http://www.humangeographies.org.ro/volume-16-issue-2-2022/1623-abstract. Zugegriffen: 8. Dezember 2022.

Kühne, O. & Koegst, L. (2022). Cartographic Representations of Coastal Land Loss in Louisiana. An Investigation Based on Deviant Cartographies. *KN – Journal of Cartography and Geographic Information 2022* (4), 1–12. https://doi.org/10.1007/s42489-022-00120-8

Kühne, O. & Berr, K. (2024). Romantik und Landschaft. In O. Kühne, F. Weber, K. Berr & C. Jenal (Hrsg.), *Handbuch Landschaft* (2. Aufl., 77–93). Wiesbaden: Springer VS.

Kühne, O., Edler, D. & Jenal, C. (2021). The Abstraction of an Idealization: Cartographic Representations of Model Railroads. Die Abstraktion der Idealisierung – über kartographische Repräsentationen von Modellbahnlandschaften. *KN – Journal of Cartography and Geographic Information 71* (2), 207–217. https://doi.org/10.1007/s42489-020-00064-x

Kühne, O., Edler, D. & Jenal, C. (2022). The cartographic representation of model railroad landscapes –theoretical considerations and empirical results from model railroad-related literature. In D. Edler, O. Kühne & C. Jenal (Hrsg.), *The Social Construction of Landscapes in Games* (S. 127–148). Wiesbaden: Springer.

Kühne, O., Berr, K. & Koegst, L. (2023a). Contingency and Landscape. Basic Considerations on Graphic and Cartographic Representations in Recourse to the Concept of Inverse Landscapes as a Contribution to Deviant Cartographies with Examples on Louisiana. *KN – Journal of Cartography and Geographic Information,* 1–12. https://doi.org/10.1007/s42489-023-00145-7

Kühne, O., Berr, K. & Koegst, L. (2023b). Contingency and Landscape: Basic Considerations on Graphic and Cartographic Representations in Recourse to the Concept of Inverse Landscapes as a Contribution to Deviant Cartographies with Examples on Louisiana. *KN – Journal of Cartography and Geographic Information 73,* 225–236. https://doi.org/10.1007/s42489-023-00145-7

Kühne, O., Koegst, L. & Berr, K. (2024). *Oilscapes of Louisiana – Neopragmatic Reflections on the Ambivalent Aesthetics of Landscape Constructions.* Wiesbaden: Springer Fachmedien.

Merlingen, M. (1999). Die Relativität von Wahrheit dargestellt am Beispiel der Entstehungsgeschichte der Wirtschafts- und Währungsunion. Ein Beitrag zur Integrationsforschung aus der Sicht des epistemologischen Konstruktivismus. *Zeitschrift für Internationale Beziehungen 6* (1), 93–128. https://www.jstor.org/stable/40843843?casa_token=r0yb2oryx8uaaaaa:skdrzv4h20s_xejc6fwnpvpjqo-cirazlchcsnfako1wm_dq4ab0ezhzppzrrpkbii07z3fqs5oge7pmkccdjwayoau--8fs6hbpjwdcucvws6scssaj. Zugegriffen: 13. November 2023.

Monmonier, M. (1993). *Mapping It Out. Expository Cartography for the Humanities and Social Sciences.* Chicago: University of Chicago Press.

Paris, R. (2005). *Normale Macht. Soziologische Essays.* Konstanz: UVK.

Ricker, B. (2020). Cartography. A compendium of design thinking for mapmakers. *International Journal of Cartography 6* (2), 263–266. https://doi.org/10.1080/23729333.2020.1715547

Rorty, R. (1991). *Essays on Heidegger and others* (Philosophical papers, v. 2). Cambridge: Cambridge University Press.

Rorty, R. (1997 [1989]. *Contingency, Irony, and Solidarity.* Cambridge: Cambridge University Press.

Rorty, R. (2003). *Wahrheit und Fortschritt.* Frankfurt (Main): Suhrkamp.

Rorty, R. (2019). *Philosophie als Kulturpolitik* (Suhrkamp Taschenbuch Wissenschaft, Bd. 2289, Erste Auflage). Berlin: Suhrkamp.

Ruhrgebietssprache. (2013, 12. April). Pillepalle. https://www.ruhrgebietssprache.de/lexikon/pillepalle.html. Zugegriffen: 30. Juli 2022.

Schenk, W. (2011). *Historische Geographie* (Geowissen kompakt). Darmstadt: WBG.

Schiewe, J. (2022). *Kartographie. Visualisierung georäumlicher Daten.* Berlin: Springer Spektrum.

Shklar, J. N. (2020). Der Liberalismus der Furcht. In H. Bajohr (Hrsg.), *Judith N. Shklar. Der Liberalismus der Furcht* (3. Aufl., S. 26–66). Berlin: Matthes & Seitz.

Topper, K. (1995). Richard Rorty, Liberalism and the Politics of Redescription. *American Political Science Review 89* (4), 954–965. https://doi.org/10.2307/2082520

Weber, M. (1976 [1922]. *Wirtschaft und Gesellschaft. Grundriß der verstehenden Soziologie.* Tübingen: Mohr Siebeck.

Wood, D. (2003). Cartography is Dead (Thank God!). *Cartographic Perspectives* (45), 4–7. https://doi.org/10.14714/cp45.497

Wood, D. & Krygier, J. (2009). Cartography. Critical Cartography. http://makingmaps.owu.edu/elsevier_geog_criticalcartography.pdf. Zugegriffen: 25. März 2024.

Chapter 4
(Self)criticism- an Assessment of the Situation

The term 'critique', which comes from the ancient Greek (kritiké [techné]) and can be translated as "[art of] judgment", already denotes "the ability to distinguish and the power of judgment" with Plato and Aristotle (Mittelstraß 2004, p. 498). In Socratic maieutics (from the Greek word for 'midwifery'), this skill (Greek: techné) of discernment serves to distinguish between the true and the false; criticism can therefore also be defined as a method of verifying truthful claims as well as proving false arguments. In modern times, Kant postulated that "systematic thinking (esprit philosophique) must prove itself as critical thinking (esprit critique)" (Mittelstraß 2004, p. 498) whereby Kant understands the paradigm of such critical thinking not as the uncovering of errors, but as a critique of pure reason as a "propaedeutic" to a "system of pure reason", which "does not lead to the expansion, but only to the purification of our reason" (Kant 1959 [1781], B 24). This purification consists in the regressive analysis or reconstruction of the efficiency and exaggerated claims of the "faculty of reason in general" (Kant 1959 [1781], A XII) both as the "destruction of unfounded orientations" and as the "construction of justified orientations" (Mittelstraß 2004, p. 499). This view of philosophical criticism can be found, for example, in critical rationalism (Andersson 1988; Popper 1989), partly in critical theory (Dubiel 1992; Habermas 1973; Horkheimer 1977 [1937], 1982; Horkheimer und Adorno 1969), as well as in constructivist philosophy of science, which at the same time is characterized as a 'critique of science' (Janich et al. 1974).

At this point, we agree with Kant's well-known assessment: "The critical path alone is still open" (Kant 1959 [1781], B 884)—but not in the sense of a one-sided understanding of critique in the traditional line of (neo)Marxist thought with its desire to change society through politics (cf. Kühne and Berr 2021). This perspective tends to lead to the dedifferentiation of science and politics. Instead, we stand in the tradition of the Enlightenment period with the separation of science and politics, notably, and the rest of (non-scientific) society in general. From the various competing theories, methods, and procedures, the most suitable is to be determined for the time being. As a basis for this positioning, we rely on relevant works by Karl Popper (for a general

O. Kühne et al., *Neopragmatism - Inverse Landscape - (Carto)graphic Representation*, Neopragmatic Horizontal Geographies 2, https://doi.org/10.1007/978-3-031-89402-2_4

introduction: Alt 1995; Boyer 2017 [1994]; Corvi 2005 [1997]; Franco 2019; Zimmer und Morgenstern 2015) and his student Ralf Dahrendorf (for an introduction to his works: Kühne 2017; Kühne und Leonardi 2020; Leonardi 2014). The relationship between dynamis and energeia (see Sect. 2.4), this means neither obscuring the realm of possibility (dynamis) through premature (or even final) determinations of reality nor reducing the realm of reality (energeia) by excluding contingent or inverse possibilities or limiting it to a few 'realizations'.

The standards of scientific honesty, which in the sense described does not claim any dogmatic authority of interpretation, also include a (self-)critical assessment of one's own claims, the claimed framework, theories, and methodologies as well as the purposes and objectives of the analyses or investigations performed. To identify the position from which such (self-)criticism will take place, and to reveal the foundations of our own academic approach to the topic for potential criticism, it is first necessary to outline in more detail which forms of criticism we wish to claim.

A cursory look at the history and forms of prospective concepts and understandings of critique initially shows different depths of intervention of critique. Immanuel Kant (1959 [1781]) understood critique, as already mentioned, as a self-examination of reason, with which the limits (Greek: krínein, to distinguish, to draw boundaries) of the scope and validity of its own epistemic, moral, and aesthetic judgments are to be defined. Representatives of the 'critical theory' of the 'Frankfurt School', drawing on Karl Marx and Sigmund Freud, understand critique as the claim to uncover structures of power and domination in society (1969). For "the facts that the senses bring to us" are "socially preformed in two ways: through the historical character of the perceived object and the historical character of the perceiving organ" (Horkheimer 1977 [1937], p. 17). This form of critique also combines socio-critical analysis with a normative claim, insofar as it does not want to stop at descriptions but strives for fundamental social change. Therefore, "nowhere does it aim merely at the increase of knowledge as such, but at the emancipation of man from enslaving conditions" (Horkheimer 1977 [1937], p. 263). Michel Foucault (2012 [1985]) aims to analyze power relationships with his form of critique. He rejects the idea of a centered, hierarchically structured power structure and instead assumes that power is inherent in every social relationship. Similar critical basic assumptions can be found in discourse theory and 'more-than-representational approaches' (such as assemblage and actor-network theory), which also focus on the investigation of power relationships, especially as involves the state and economy (Glasze et al. 2021; Mattissek und Wiertz 2014; Miggelbrink 2014).

The general concept of critique can be defined more precisely if two aspects are considered regarding the research literature. First, every critique requires a criterion for its critique; second, the fundamental distinction between internal (or immanent) and external (or transcendent) critique must be taken into account (cf. exemplarily Stederoth 2011). As far as the standard of critique is concerned, we neopragmatically advocate deriving this standard from the respective concrete *purpose* of either research or critique. If it is also taken into account that the criterion of criticism must be related to the standards of the criticized (Stederoth 2011, p. 1354) then the extent

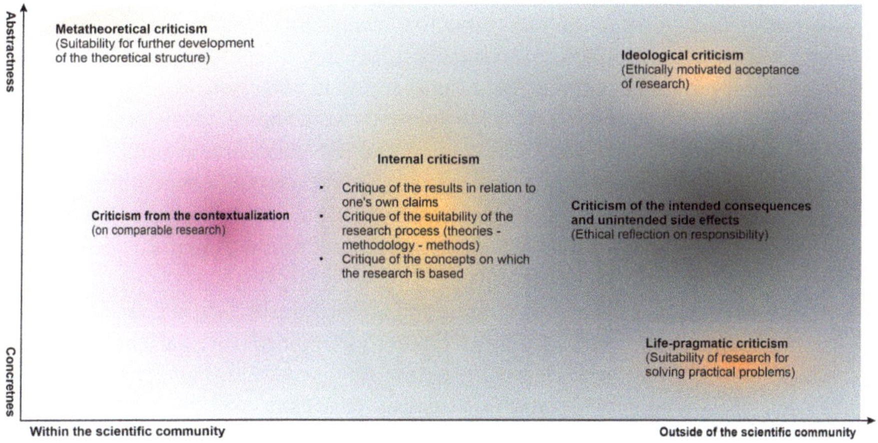

Fig. 4.1 Different types of criticism, localized in the dimensions of abstractness and concreteness as well as being internal and external to science (Kühne et al. 2023, S. 39)

to which internal or external criticism can (or wants to) engage with the standards of the criticized must be taken into account.

Internal or immanent criticism can engage with the standards of that which is criticized and examine whether, and to what extent, which being criticized does justice to these own standards. External or 'transcendent' criticism, on the other hand, only judges the criticized by the standards of its own criticism. The latter aspect points to an inherent danger of external criticism, which Ralf Dahrendorf, Odo Marquard, and Hillary Putnam, for example, have explicitly pointed out: namely, the danger of a totalization of criticism that can no longer acknowledge any standards other than its own. Dahrendorf emphasizes that academics must guard against two extreme positions: "the Scylla of a rather conservative insistence on inviolable institutions, and the Charybdis of a completely unrestrained reformism, to which all reliability ultimately falls victim" (Dahrendorf 1994, p. 176). Similarly, Marquard warns against an "absolute critique of time" (Marquard 2007, p. 117) in whose totalizing framework people can be instructed or even suppressed through revolutionary or pedagogical measures. Putnam warns against a totalizing deconstruction that ultimately "boils down to the fact that *everything* can be 'deconstructed'", so that this form of deconstruction "has no morality" (Putnam 1997, p. 252 emphasis in the original).

A distinction can be made between six typical forms of criticism, which can be distinguished on the basis of the polarity of abstractness and concreteness as well as the polarity of internal and external criticism (Fig. 4.1), can be defined in more detail. These forms of criticism have relatively large reference areas, which are indicated by corresponding color gradients. We distinguish between the following forms of criticism (cf. Kühne et al. 2023, p. 39–41):

1. As the most elementary form of criticism, *internal criticism* (cf. Stederoth 2011; Boltanski 2010; Jaeggi 2009; Stahl 2013). The critical reference point is the

framework and purpose of the respective research; the question is therefore whether the self-formulated claim is truly fulfilled. Criticism of the suitability of the research process is central; for example, it is necessary to examine whether theories, concepts, methodologies, and methods were appropriately chosen, sufficiently formulated, contextualized clearly, and led to suitable results. This self-referential criticism can be performed as a reflection on one's own approach, but it can also be brought to the pursued research from outside (e.g., by reviewers or critics). Internal criticism tends to take place within the scientific community but can also be performed externally.

2. Another typical and proven form of internal scientific criticism is *criticism based on contextualization.* The criticized research is placed in relation to similar research about the choice of theory, terminology, methodology, methods, and data selection. The consideration of relevant literature is in any case a central basis of scientific work. This form of internal scientific criticism can assume varying degrees of abstraction.
3. *Metatheoretical criticism* is also internal to science and is also very abstract. It examines whether and to what extent research is suitable for expanding the theoretical spectrum or for testing, restricting, and expanding other theoretical approaches. Of course, this form of criticism requires justification, as the metatheoretical framework (in this case neopragmatism) must be disclosed and identified to be able to understand and justify the claimed standards of criticism.
4. *Worldview criticism* resides within a context external to science and is also very abstract. It can be religiously, politically, economically, culturally, or morally based and is usually motivated by ethical convictions (in the sense of Weber 1976 [1922], 1988) and tends neither to clearly identify the basis of the criticism nor to question or justify it. Since ideological criticism is often morally charged, it tends to criticize, discredit, and pathologize both research and the researchers themselves (Grau 2017; Luhmann 1993). This form of criticism is usually not very differentiated and is based on untested generalities.
5. *Criticizing the consequences and unintended side effects* of research (Berger 2017 [1963]; Dahrendorf 1968; Popper 1963, 1992) is an ethical orientation in terms of responsibility (in the sense of Weber 1976 [1922]). It examines the responsibility of the research results; the ethical position from which a judgment is made must be clearly stated. This form of criticism can be applied at all levels of abstraction as well as within and outside the scientific community.
6. *Life-related pragmatic criticism,* on the other hand, is not very abstract, but rather concretely related to practical life challenges and strongly oriented towards the outside of science. It examines the suitability of research in relation to the solution of practical life problems and therefore has an explicit reference to the real world. The basis of the critique is hardly in need of justification here, as it is the obvious starting point of the critique: the lifeworld (pertinent: Gethmann 1991; Gethmann et al. 2011; Husserl 1954, 2008; Janich 2011; Mittelstraß 1991; Schütz und Luckmann 2003 [1975]; Waldenfels 2005; Welter 1986).

The subject of criticism in this book is the formation of the consciousness of the individual incurring conflict by means of inversion. The question is which strategies are responsible for the development or limitation of the individual's consciousness. This involves an individual who is confronted with inversions in his everyday life. The concept of consciousness that is taken as a basis here is one that, in addition to reason and understanding, also includes emotion on an equal footing and assumes an equal originality of individual, interpersonal, and environmental consciousness. With regard to 'feelings' in particular, the assumption here is that feelings are an emotional cognitive faculty that has a value-forming function (Frazzetto 2014; Schwarz 2012). According to Schwarz, "analytical[...] thinking that draws on facts and arguments" must be distinguished from "thought processes that are not fully accessible [...]. The latter are [...] called intuitive" (Schwarz und Krex 2022). These are "feeling-based judgments that seem right to you without you knowing the reason for them" (Schwarz in Schwarz und Krex 2022). An "essential feature of intuition" is that it enables the individual to "maneuver through everyday life preconsciously. Intuition tells us whether something requires our attention or not. We could not cope with everyday life if we had to analyze everything around us, always pay attention, listen everywhere and look closely. Intuition helps us to prioritize" (Schwarz und Krex 2022). This power of intuition is given to us 'out of the blue', or in other words, it is intrinsic to us. Emotion as an immediate guide is more than a mere feeling. Alongside rationality, it is a form of knowledge and, similar to cognitive memory, there is also an intuitive memory: "An unconscious decision not to walk down a certain street at night, for example, often arises from the accumulation of previous experiences. In this sense, one can [...] say that these intuitive signals take more knowledge into account than if one were to collect facts in a hurry" (Schwarz und Krex 2022). Feelings are "informative" and "we use them [...] systematically to make judgments. We are very sensitive to our feelings, but we are insensitive to where they come from" (Schwarz und Krex 2022). According to Schwarz, it is instructive to affect the emotional life of the individual in such a way that "mistrust" (Schwarz und Krex 2022)—i.e., irritation—arises, because according to Schwarz, individuals are much more critical in this emotional state. This is precisely what needs to be illuminated by means of an inversion as an instrument of consciousness formation (on the enormous and sometimes disparate scope of the definitions of consciousness, in general, Mutschler 2017). The aim of the formation of the individual's consciousness envisaged here is a self-educated insight into and a self-determined attitude towards the conflict—or, in Schwarz's words, a "self-assessment" (Schwarz und Krex 2022)—as well as the willingness to engage in a solution-oriented manner.

In conversational practice, criticism (like irony) aims to rethink a point of view through uncertainty. The prerequisite for a constructive dynamic is that no reflexive (emotionally) blocking attitude is created, but that an open process of reflection can be initiated. Since criticism carries the risk of creating a feeling of vulnerability and/or experiencing a loss of meaning, the tendency towards a reflexive self-protective attitude is obvious and "the person [sic!] who wants to criticize must skillfully prevent this self-protective attitude if he (a) wants to reach the addressee and (b) wants to convince the audience of his own seriousness and encourage them to adopt his attitude. Polemical or irrational criticism often fails to achieve this goal in such discourses" (Schwerz-Friesel et al. 2012, p. 239–240). Criticism thus has the special feature that not only the understanding of the content (factual level) is achieved, but at the same time a fundamental acceptance of the person must be noticeable. Rogers (2005) speaks of three basic characteristics for a growth-promoting discussion framework on the sending side: empathy, congruence, and positive emotional appreciation. On the receiving side, Popper believes that the basis for a responsibility-based communication attitude is that growth only works if the interacting partners are challenged to contradict each other and are "more interested in learning than in being right" (Popper 1996, p. 169). The prerequisite for criticism is therefore that the content level is separated from the level of their relationship (Schulz von Thun 2019).

References

Alt, J. A. (1995). *Karl R. Popper* (2. Aufl.). Frankfurt: Campus.

Andersson, G. B. J. (1988). *Kritik und Wissenschaftsgeschichte. Kuhns, Lakatos' und Feyerabends Kritik des kritischen Rationalismus* (Die Einheit der Gesellschaftswissenschaften, Band 54). Tübingen: Mohr Siebeck.

Berger, P. L. (2017 [1963]. *Einladung zur Soziologie. Eine humanistische Perspektive* (2., ergänzte Auflage). Konstanz: UVK; UVK/Lucius. (Originalarbeit erschienen 1963).

Boltanski, L. (2010). *Soziologie und Sozialkritik*. Berlin: Suhrkamp (Frankfurter Adorno-Vorlesungen 2008).

Boyer, A. (2017 [1994]. *Introduction à la lecture de Karl Popper*. Paris: Éditions Rue d'Ulm.

Corvi, R. (2005 [1997]. *An introduction to the thought of Karl Popper*. London: Routledge.

Dahrendorf, R. (1968). *Pfade aus Utopia. Arbeiten zur Theorie und Methode der Soziologie*. München: Piper.

Dahrendorf, R. (1994). *Der moderne soziale Konflikt. Essay zur Politik der Freiheit*. München: dtv.

Dubiel, H. (1992). *Kritische Theorie der Gesellschaft. Eine einführende Rekonstruktion von den Anfängen im Horkheimer-Kreis bis Habermas*. Weinheim: Beltz Juventa.

Foucault, M. (2012 [1985]. *Discipline and Punish: The Birth of the Prison*. New York: Knopf Doubleday Publishing Group.

Franco, G. (Hrsg.). (2019). *Handbuch Karl Popper* (Living reference work). Wiesbaden: Springer Reference Geisteswissenschaften.

Frazzetto, G. (2014). *Der Gefühlscode. Die Entschlüsselung unserer Emotionen*. München: Hanser.

Gethmann, C. F. (Hrsg.). (1991). *Lebenswelt und Wissenschaft. Studien zum Verhältnis von Phänomenologie und Wissenschaftstheorie* (Neuzeit und Gegenwart, Bd. 1). Bonn: Bouvier.

Gethmann, C. F., Bottek, J. C. & Hiekel, S. (Hrsg.). (2011). *Lebenswelt und Wissenschaft. XXI. Deutscher Kongreß für Philosophie 15. – 19. September 2008 an der Universität Duisburg – Essen*. Hamburg: Felix Meiner Verlag (Kolloquienbeiträge).

Glasze, G., Bittner, C., Michel, B. & Strüver, A. (2021). Ein diskurstheoretisch informierter Blick auf Karten und Kartographie. In G. Glasze & A. Mattissek (Hrsg.), *Handbuch Diskurs und Raum. Theorien und Methoden für die Humangeographie sowie die sozial- und kulturwissenschaftliche Raumforschung* (S. 405–416). Bielefeld: transcript.

Grau, A. (2017). *Hypermoral. Die neue Lust an der Empörung* (2. Aufl.). München: Claudius.

Habermas, J. (1973). *Erkenntnis und Interesse. Mit einem neuen Nachwort*. Frankfurt (Main): Suhrkamp.

Horkheimer, M. (1977 [1937]. *Traditionelle und kritische Theorie. Fünf Aufsätze*. Frankfurt (Main): Fischer.

Horkheimer, M. (1982). *Critical Theory. Selected Essays*. New York: continuum.

Horkheimer, M. & Adorno, T. W. (1969). *Dialektik der Auffilärung. Philosophische Fragmente*. Frankfurt (Main): Fischer.

Husserl, E. (1954). *Die Krisis der europäischen Wissenschaften und die transzendentale Phänomenologie. Eine Einleitung in die phänomenologische Philosophie* (Husserliana, Bd. 6). Den Haag: Martinus Nijhoff (Herausgegeben von Walter Biemel).

Husserl, E. (2008). *Die Lebenswelt. Auslegungen der vorgegebenen Welt und ihrer Konstitution*. Dordrecht: Springer (Texte aus dem Nachlass (1916–1937)).

Jaeggi, R. (2009). Was ist Ideologiekritik? In R. Jaeggi & T. Wesche (Hrsg.), *Was ist Kritik?* (S. 266–298). Frankfurt (Main): Suhrkamp.

Janich, P. (2011). Handwerk und Mundwerk. Lebenswelt als Ursprung wissenschaftlicher Rationalität. In C. F. Gethmann, J. C. Bottek & S. Hiekel (Hrsg.), *Lebenswelt und Wissenschaft. XXI. Deutscher Kongreß für Philosophie 15. – 19. September 2008 an der Universität Duisburg – Essen*. Kolloquienbeiträge (S. 678–691). Hamburg: Felix Meiner Verlag.

Janich, P., Kambartel, F. & Mittelstrass, J. (1974). *Wissenschaftstheorie als Wissenschaftskritik*. Frankfurt (Main): Aspekte Verlag.

Kant, I. (1959 [1781]. *Kritik der reinen Vernunft*. Hamburg: Felix Meiner Verlag.

Kühne, O. (2017). *Zur Aktualität von Ralf Dahrendorf. Einführung in sein Werk* (Aktuelle und klassische Sozial- und Kulturwissenschaftlerlinnen). Wiesbaden: Springer VS.

Kühne, O. & Leonardi, L. (2020). *Ralf Dahrendorf. Between Social Theory and Political Practice*. London: Palgrave Macmillan.

Kühne, O. & Berr, K. (2021). *Wissenschaft, Raum, Gesellschaft. Eine Einführung zur sozialen Erzeugung von Wissen*. Wiesbaden: Springer VS.

Kühne, O., Berr, K. & Lohmann, P. (2023). *Landschaft zwischen Philosophie und Sozialwissenschaften. Eine Kritik*. Wiesbaden: Springer Fachmedien; Springer VS.

Leonardi, L. (2014). *Introduzione a Dahrendorf* (Maestri del Novecento, Bd. 20). Roma: Editori Laterza.

Luhmann, N. (1993). Die Moral des Risikos und das Risiko der Moral. In G. Bechmann (Hrsg.), *Risiko und Gesellschaft. Grundlagen und Ergebnisse interdisziplinärer Risikoforschung* (S. 327–338). Opladen: Westdeutscher Verlag.

Marquard, O. (2007). Freiheit und Pluralität. In *Skepsis in der Moderne. Philosophische Studien* (S. 109–123). Stuttgart: Reclam.

Mattissek, A. & Wiertz, T. (2014). Materialität und Macht im Spiegel der Assemblage-Theorie: Erkundungen am Beispiel der Waldpolitik in Thailand. *Geographica Helvetica 69* (3), 157–169.

Miggelbrink, J. (2014). Diskurs, Machttechnik, Assemblage. Neue Impulse für eine regionalgeographische Forschung. *Geographische Zeitschrift 102* (1), 25–40.

Mittelstraß, J. (1991). Das lebensweltliche Apriori: Paul Lorenzen zum 70. Geburtstag. In C. F. Gethmann (Hrsg.), *Lebenswelt und Wissenschaft* (S. 114–142). Bonn.

Mittelstraß, J. (Hrsg.). (2004). *Enzyklopädie Philosophie und Wissenschaftstheorie* (2, H-O, unveränderte Sonderausgabe). Stuttgart: J.B. Metzler.

Mutschler, H.-D. (2017). Was ist Bewusstsein? *Theologie und Philosophie 92* (1), 1–37.

Popper, K. R. (1963). *Conjectures and refutations. The growth of scientific knowledge*. London: Routledge & Kegan.

Popper, K. R. (1989). *Logik der Forschung*. Tübingen: Mohr Siebeck.

Popper, K. R. (1992). *Die offene Gesellschaft und ihre Feinde. Falsche Propheten – Hegel Marx und die Folgen* (Bd. 2, 7. Aufl., 2 Bände). Tübingen: J.C.B. Mohr. (Originalarbeit erschienen 1945).

Popper, K. R. (1996). *Alles Leben ist Problemlösen. Über Erkenntnis, Geschichte und Politik*. München: Piper.

Putnam, H. (1997). *Für eine Erneuerung der Philosophie* (Universal-Bibliothek, Bd. 9660). Stuttgart: Reclam.

Rogers, C. R. (2005). *Die klientenzentrierte Gesprächspsychotherapie. Client-Centered Therapy* (17. Aufl.). Frankfurt: Fischer.

Schulz von Thun, F. (2019). *Miteinander reden. 1–4: Störungen und Klärungen/Stile, Werte und Persönlichkeitsentwicklung/Das „Innere Team" und situationsgerechte Kommunikation/Fragen und Antworten.* Hamburg: Rowohlt Taschenbuch.

Schütz, A. & Luckmann, T. (2003 [1975]. *Strukturen der Lebenswelt.* Konstanz: UTB.

Schwarz, N. (2012). Feelings-as-Information Theory. In P. van Lange, A. Kruglanski & E. Higgins (Hrsg.), *Handbook of Theories of Social Psychology* (S. 289–308). London: SAGE.

Schwarz, N. & Krex, A. (2022). "Riecht es nach Fisch, sind die Menschen misstrauischer". Interview mit Norbert Schwarz in Zeit online vom 22. April 2022. https://www.zeit.de/entdecken/2022-04/intuition-psychologie-misstrauen-sozialpsychologe-norbert-schwarz/komplettansicht. Zugegriffen: 29. Juni 2024.

Schwerz-Friesel, M., Marx, K. & Damisch, S. (2012). Persuasive Strategien der affektiven Verunsicherung im aktuellen Diskurs: Ironisieren, Kritisieren und Beleidigen in öffentlichen Streitgesprächen. In I. Pohl & H. Ehrhardt (Hrsg.), *Sprache und Emotion in öffentlicher Kommunikation* (S. 227–254). Frankfurt (Main): Peter Lang.

Stahl, T. (2013). *Immanente Kritik. Elemente einer Theorie sozialer Praktiken.* Frankfurt (Main):Campus.

Stederoth, D. (2011). Kritik. In P. Kolmer & A. G. Wildfeuer (Hrsg.), *Neues Handbuch philosophischer Grundbegriffe* (Bd. 2, S. 1346–1357). Freiburg (Breisgau): Alber.

Waldenfels, B. (2005). *In den Netzen der Lebenswelt* (3. Aufl.). Frankfurt (Main): Suhrkamp.

Weber, M. (1976 [1922]. *Wirtschaft und Gesellschaft. Grundriß der verstehenden Soziologie.* Tübingen: Mohr Siebeck.

Weber, M. (1988). *Gesammelte Politische Schriften von Max Weber* (5. Aufl.). Tübingen: Mohr Siebeck.

Welter, R. (1986). *Der Begriff der Lebenswelt. Theorien vortheoretischer Erfahrungswelt* (Übergänge, Bd. 14). München: Fink.

Zimmer, R. & Morgenstern, M. (2015). *Karl R. Popper. Eine Einführung in Leben und Werk* (2., durchgesehene und ergänzte Auflage). Tübingen: Mohr Siebeck.

Chapter 5
Critique of the Concept of Inverse Landscapes and Their Representations

Critiques can be understood as the construction of analytical categories that are intended to facilitate an approach. This does not constitute an ontological statement about the field of critique and certainly not a statement about an 'essence'. In this respect, the category system is a tool to address questions that examine the theory of inverse landscapes with the aim of increasing the usefulness of the theory. Following this idea of tools and usefulness, we will use the category system in the following, sometimes more explicitly and sometimes more implicitly. Similarly, we will use the category system less so when not all aspects of the criticism are affected or if presentations according to this system would be too schematic and thus hinder rather than help the understanding of the considerations. The criticism presented here is certainly not complete. Additional aspects of the criticism can be brought forward from other perspectives (including ours, conceivably). The critique presented here is intended—in the neopragmatist sense—as a tool for sharpening the usefulness of neopragmatist landscape theory, including its concept of inverse landscapes. Critique, as a tool here, means that the goal is to sharpen the specifics of the theoretical vocabulary used to communicate aspects of a neopragmatist landscape, to clarify where the limits and boundaries of this vocabulary are to be located (which must remain excluded, presently), and at which points critique can lead to a further development of the approach.

We will then address the criticisms made in this chapter in the following Chapter 6 ("Valuing criticisms: extending and specifying inverse landscapes and their representations"). There we will go through them point by point, reject criticism, accept criticism, and adapt the concept of neopragmatist inverse landscapes (also in their components) or use the criticisms to clarify where the limits of the concept lie. These two chapters are organized in such a way that it is easy to find the explanations on how to deal with the criticisms. This means that if the handling of each criticism is to be read directly, Sect. 5."x" can be read first, followed by the answer in the correlating Sect. 6."x" without any problems. Of course, it is also possible—and this

O. Kühne et al., *Neopragmatism - Inverse Landscape - (Carto)graphic Representation*, Neopragmatic Horizontal Geographies 2, https://doi.org/10.1007/978-3-031-89402-2_5

is suggested by the structure of the work—to read the 'concentrated criticism' first and then read the way it is engaged in a coherent form.

5.1 Critique 1: Prerequisites for a Neopragmatic Landscape Science

Pertinent to the prerequisites for neopragmatic landscape science, the high degree of complexity involving the research process can be emphasized *internally*. A comprehensive knowledge of theories, methods, data, researcher perspectives, and methods of representation is a prerequisite for a neopragmatist approach to landscape. In the *context* of essentialist and positivist science, neopragmatism is *criticized* as relativistic. It has (like other [social] constructivist approaches) abandoned the idea of an 'objective truth' (see in this context: Kneer 2009; Knoblauch 2016; Krohn 2000). Science is not adjudged as a neutral 'observer of the world', but as a tool for generating useful knowledge. With the focus on linguistic theory, there is the problem of constructivist perspectives (Holmén 1995) being able to take appropriate account of the material. From a *metatheoretical* perspective, it can first be criticized that despite all sensitivity to contingency and all awareness of complexity, there is a tendency to reduce complexity and tame contingency. This also becomes clear in the categorical separation of public and private spheres, which creates a dichotomization that neopragmatism wanted to overcome with its pan-relationalism. From a metalevel perspective, it can also be criticized that neopragmatism functions in current landscape research as a metatheoretical regulation mechanism for dealing with vocabularies, while the pan-relationalist potentials have not yet been exhausted. *World-view* criticism of neopragmatism can be expressed essentially from a collectivist and utopian perspective, because neopragmatism, with its focus on practical utility, is not suitable for the creation of 'utopias' beyond the 'meta-utopia' of an inclusive society with the expansion of the 'we'. With this basic ideological position of expanding the 'we' and avoiding pain, neopragmatism is geared towards the expansion of a liberal democratic society (with which neither a communist nor a nationally unified society, nor a 'theocracy' can be founded). Referring to the *criticism of the consequences and unintended side effects*, it can be stated that utilitarian thinking and falsificationism are hardly suitable as an integrative idea for the future of a society, making it difficult to achieve a 'theocracy'. The focus on procedural justice also fails to account for the unequal distribution of life chances. Not least, this can have the consequence that the expansion of the cultural, social, and political 'we' is accompanied by increasing economic polarization. The *life-related pragmatic* criticism, in turn, relates to the usefulness of the results of neopragmatism, and the comprehensibility of the results is also made more difficult by the high degree of complexity resulting from the diversity of theories, methods, data, et cetera. At the same time, socialization conditions and educational careers (Bourdieu 1987 [1979]), as well as discipline-specific knowledge, mean that the conditions of access to knowledge and discourse options are subject to milieu-specific and individual-specific hurdles and are not equally distributed among the members of society.

5.2 Critique 2: Neopragmatist Understandings of Landscape

In terms of an *internal criticism*, a neopragmatist understanding of landscape can be assessed a high degree of complexity. This is notably expressed through the structuring into Landscape 1, Landscape 2, and Landscape 3, as well as through the differentiated construction of landscape in Mode A, Mode B, and Mode C. This high degree of complexity presents neopragmatist landscape researchers with the challenge of clarifying the usefulness of the approach. The connection between pragmatism's utilitarian thinking as well as contingency awareness and creation (also) require explanation in relation to landscape research. From the *contextualization*—with (other) theoretical positions—it can be criticized that evolutionary theories (for example: Appleton 1975, 1984; Kaplan und Kaplan 1989; Orians 1980, 1986) are not included in the context of landscape theories considered in the theoretical triangulation. From the perspective of critical landscape research, in turn, neopragmatist landscape research does not attach sufficient importance to the issue of power.

Metatheoretical criticism is directed at the quasi-ontological theory of the three landscapes on which the (here, an elaborated understanding of) neopragmatist landscape theory is based. The theory of the three landscapes focuses (as outcome of the underlying three world theory) strongly on the significance of individual consciousness, especially its central position of mediation between Landscape 3 and Landscape 1. The significance of intersubjective exchange is only barely considered, at the utmost. Inverse landscapes do not arise solely from the engagement with Landscape 1 or Landscape 3, but through intersubjective engagement with different understandings of landscape. On the other hand, the question must be asked whether the categorization into three landscapes does not run counter to the pan-relationalist understanding of neopragmatism, as it assumes a (more or less) fixed structuring upon the world of that which is tangible, the individual consciousness, and the respective cultural contents. Ultimately, the *ideological* criticism of neopragmatist landscape theory represents a landscape specification of the general criticism of neopragmatism. From the perspective of collectivist theories, the high conceptual significance of Landscape 2 in comparison to Landscape 3, from the perspective of theories that see the tangible as the constitutive basis of landscape, can be criticized in particular. Criticism of *consequences and unintended side effects*: Contingency and complexity require the willingness and ability not only to endure these, but moreover, if they are to be made fruitful in neopragmatist terms, also to deal with these constructively. This certainly requires effort, as it runs counter to the reflexive tendency to draw quick cognitive conclusions and to focus on confirming opinions and expectations that have formed already. If there is no willingness to make such efforts, this often leads to a desire for unambiguity, for example, in the form of the 'one (known) truth'. This makes neopragmatist landscape research an unsuitable basis for a practice that relies on unambiguous solutions to complex problems. From the perspective of *life-related pragmatic* critique, the richness of presuppositions of the neopragmatist understanding of landscape can be questioned, in turn. This richness of presuppositions makes neopragmatist landscape concepts complex and often lacking accessibility intuitively.

5.3 Critique 3: The Concept of Inverse Landscapes

When approaching the concept of inverse landscape *with an internally critical eye*, it can be noted that the manifestation of inverse landscapes can be understood as contradictory to the idea of contingency of neopragmatism. Moreover, against the background of the more-than-representational claim of neopragmatism, inverse landscapes can be comprehended as very strongly oriented towards representation, even if contingent landscapes are presented. As concerns *contextualization*, there is a need for greater consideration of temporal, spatial, and cultural contexts. From this critical perspective, it can also be notable that the concept of inverse landscapes assumes a 'complete inversion' and does not include the possibility of an internal combination of contingent and non- contingent components within inverse landscapes. From a *metatheoretical* perspective, we can agree with this criticism of the *contextualization*, which, however, is linked to a mandate for adaptation: Given the pan-relationalist orientation of neopragmatism underlying the concept of inverse landscape with its sensibility towards hybridity, a binary encoding of inversion and non-inversion cannot be comprehended. *Worldview* criticism of the concept of inverse landscape is directed from an essentialist perspective, that the creation of inverse landscapes only obscures the search for the recognition of the 'essence of landscape'. From a critical theory perspective, it can be argued that the concept, with its focus on the expansion of contingent approaches, merely leads to a dilution of the power of tangible landscape utopias. Criticism of the *consequences and unintended side-effects* of the concept of inverse landscapes raises the problem that exclusivist inverse landscapes could also be created, which would therefore be associated with a loss of contingency. Such exclusivist inverse landscapes could even be used to reshape Space 1 interpreted as landscape in a contingency- reducing manner. From the perspective *of life-related pragmatic* criticism, a perception that is well removed from life-as-lived can be emphasized, which in turn (consciously or unconsciously) impedes any further examination of contingency.

5.4 Critique 4: Inverse Landscapes as Contingency—Potentials and Challenges

When approaching the concept of inverse landscape *with a self-critical eye*, it can be noted that the concretization of inverse landscapes can be understood as contrary to the contingency idea of neopragmatism because of the "need for completion" (Lewin 1927, p. 85). Moreover, against the background of the more-than-representational claim of neopragmatism, inverse landscapes are perceivable as very strongly oriented towards representation, even if contingent landscapes are exhibited. As concerns *contextualization*, there is a need for greater consideration of temporal, spatial, and cultural contexts. From this critical perspective, it can also be noted that the concept of inverse landscapes assumes a 'complete inversion' and does not account for the possibility of an internal combination of contingent and non-contingent components within inverse landscapes. From a *metatheoretical* perspective, we can agree with this criticism from the *contextualization*, which is, however, linked to a mandate for adaptation: Given the pan-relationalist orientation of neopragmatism underlying the concept of inverse landscape with its sensibility of hybridity, a binary coding of inversion and non-inversion cannot be comprehended. *Worldview* criticism of the concept of inverse landscape is directed from an essentialist perspective; that the creation of inverse landscapes only obscures the search for the recognition of the 'essence of landscape'. From a critical theory perspective—following on from the discussion in Sect. 5.3—the concept of inverse landscapes, with its focus on the expansion of contingent access, only leads to a dilution of the power of concrete landscape utopias and, with the help of (carto)graphic representations, a kind of escapist opportunity is created by means of an appealing design, which only distracts from the central social problems. From the discussion in Sect. 5.3, the argument can be raised that the (carto)graphic design of inverse landscapes, in particular, could also create exclusivist inverse landscapes. A loss of contingency is thereby associated with this design. Representations of exclusivist inverse landscapes created in this way could form a basis for reshaping Space 1 in a way that reduces contingency. From the perspective of a *life-related pragmatic* critique, a conception of inverse landscapes that is far removed from the practice of life can be criticized (see Sect. 5.3), which is also reflected in their (carto)graphic representation.

5.5 Critique 5: Inverse Landscapes as Ironic (Self-) attitude—Potentials and Challenges

An *internal critique* of the representation of inverse landscapes is performed based on the question of whether the self-formulated claim to be an expression of an ironic (self-)attitude and, therefore, to promote an awareness towards contingency, plus thinking in terms of possibilities, can be fulfilled. Here it must be (self-)critically stated that endangering the fulfillment of this claim are various internal framework conditions. First and foremost is the duality of irony (see Sect. 2.7), via enabling possibilities of experiencing contingency and thus freedom to view the world, and contrastingly, evoking reactance if the EGO or ALTER ('double contingency') is incapable of irony. The ability to use irony requires that the vocabulary and representations used are understandable by the recipients. This addresses a contradiction between production (ironic texts and representations) and reception (understanding and accepting the texts and representations), as classically discussed in aesthetics as a philosophy of art. This means that even if irony creates spaces of possibility, the essential prerequisites for the production and reception of irony must be considered, but these cannot be assumed to be present in recipients. Nor is it guaranteed that the representation of inverse landscapes will necessarily lead to an expansion of the space of contingency. For this to be possible, a disturbance, a maintaining of possibilities must be permitted within a picture. However, this is not an a priori theoretical question, but an empirical one that cannot be answered in advance (of reception) in view of the described 'double contingency' of irony production and irony reception. Appropriate research into perception and reception would be required here, which could potentially provide a rule of thumb. Another problem in this context is the danger of interpretative determinism in text and image combinations. It cannot be ruled out that commenting image texts suggest an interpretation that—even as a contingent—intentionally or unintentionally determines a specific perception and interpretation for recipients. The depiction of inverse landscapes is undoubtedly a production of irony in Mode C; it is intended to increase options for experiencing contingency. However, this collides with the typical mode of reception of everyday recipients, Mode A (unless a publication is only intended to be addressed internally to other scientists) and Mode B, which can come into conflict with Mode C.

A critique from the *contextualization* can directly follow on from the internal critique and call for recipients to be more intricately involved in the research design in their contingent ability to receive ironic representations.

A *metatheoretical* critique can, for example, urge us to go beyond Rorty's conception of irony, on which we rely here. For Rorty, the ironist is limited to the private sphere and the individual. A critical objection to this is that irony unfolds its effect in an intersubjective exchange. However, as shown, this requires cultural capital, which determines the capacity for irony of EGO and ALTER. In this way, irony is also—in terms of its constitutive conditions—located in the public sphere as the sphere of intersubjective communication, understanding, and interaction. In the public sphere, the differentiation between role (a person in a public function) and individual (a 'self' even before a role is ascribed) is also an indispensable fact of successful human coexistence. At present, however, it can be observed that a dedifferentiation of role and person is taking place in current social and cultural science, as well as public communication: The person is determined via socio-cultural attributions ('old white man'), for example. A personification takes the place of a debate with argumentation, which can also take place in an ironic manner (Ng 2020; Rorty 1997; Stegemann 2018). This personification, in turn, knows not irony, because this presupposes role distance, which in turn is linked to the recognition of the differentiation between individual and role (whereby hybridizations can also occur here). A neopragmatist approach to landscapes and their representations must deal with this development. Furthermore, an inverse landscape in an artistic context could be a representation that, for example, reverses the usual conventions or expectations of landscape painting. This could mean that traditional elements of a landscape, such as the sky or vegetation, are depicted differently than would be expected.

An *ideological* critique could confront the liberal basic understanding of the concept of irony with other basic positions and criticize the fact that, in view of urgent crises such as climate change, energy, supplies of raw materials, food production, and transport infrastructures, etc., landscape conditions must not be left to irony and playing with contingencies, but must be subjected to a 'no- alternative', serious, resilient knowledge of disposition.

As involves the desired increase in contingency and the sensitivity of contingency through ironic forms of representation, a critique of the *consequences and unintended side effects* could point to the danger that this could lead to a relativization of 'realistic' representations of reality or an arbitrariness of forms of representation in general. This could discredit an earnest, serious, and realistic representation of the landscape.

Finally, a *pragmatic* critique could argue that the danger of the misunderstanding or rejection of irony is transferred to the forms of representation and that the goal of a contingency-sensitive orientation for recipients in practical life contexts is missed.

5.6 Critique 6: Emotion and Compassion Integrated into Inverse Landscape Concepts

As involves the integration of feeling and compassion into the concept of inverse landscapes, the following internal *criticism* can be raised: The "connection" between the "immediacy of the sense of self with the certainty of existence" (Frank 2002, p. 7; Lohmann 2006; Stalfort 2013) is compelling for the concept of inverse landscapes. The irritation triggered by irony is initially a perceived deviation from the known, of which one was previously certain. This reversal from the certain to the uncertain is expressed in the self-relation as a feeling (Oesterreich 2011, p. 50). As a result of this inner turning point, the individual embarks on an open path and, under certain circumstances, embarks on paths that were previously hidden, i.e., paths that they were not even aware of, because they usually adopt familiar assumptions and a proven pattern of behavior through anticipating the event that follows the inversion, which corresponds to what they have experienced so far. However, through turning it into an irritation, it becomes possible to distance oneself from traditions, break down prejudices, become more willing to take risks, and enter a free, unknown field. Emotions guide this because they are themselves a kind of knowledge (cf. Schwarz und Krex 2022). Inversion proves to be a tool for critical behavior. It strengthens autonomy and trust in oneself. As a result, it motivates self-education, as well as the ability to endure potential failure and to continue to develop oneself through what has been overcome. Inversions therefore promote the ability to have an 'agile' mind, i.e., to perceive perspectives on what happens to you via the alternative perspective not only from an internal point of view, but also to always combine these insights with the perspective of others. This is because landscapes, per se, have interpersonal prerequisites. Since this connection with oneself, the other, and the landscape have their common root in feelings (Berr et al. 2023; Kühne et al. 2023), this connection is inherent in every individual which is inseparably and irrevocably linked to them. It follows that by reflecting on the emotion(s) triggered by the inversion, the individual not only experiences irritation, but also, in understanding what irritates them, experiences their power over themselves, their relationship to others, and to the landscape. The latter is not meant in the sense of determination, but conversely as a strengthening of the ability to create (cf. Derrida 2003, p. 32) new perspectives. By means of mental activity stimulated by inversion, the individual is the originator of possibilities. The motivation and initial orientation in terms of content as to how one reacts to a conflict originally lie within oneself via feelings.

Inversions, according to the criticism from *contextualization*, cause irritation in the individual. As a result, that which is familiar, which one engages with subconsciously, becomes problematic and uncertain. Practical strategies for coping with everyday life can shift in their hierarchy from proven to less proven and become questionable themselves. The merely existing and known, as a basis for decisions in a conflict, are no longer safe guides. The individual must decide for himself, and this is likely to lead to the breaking up of familiar patterns of behavior. The answer to the solution of a conflict is pending and must be formed by the individual (cf. Schwarz 2022). The self-relationship determined by inversion proves to be a task. With this task, the individual is connected to the other, because landscape is always determined interpersonally. The task of the individual is, in one, always also the co-task of the other. The solution to the task lies in shared responsibility (Mieg 2001).

From a *metatheoretical* perspective, the following criticism can be formulated regarding the integration of emotion and compassion into the concept of inverse landscapes: Without inversion, no new view of the familiar, no breaking out of persistence are possible. Having to take a new approach to what is familiar encourages creativity and provides a horizon of possibilities where previously there was only factual, requisite reality. The side of the opposite revealed by inversion now appears as an alternative. This makes you aware of something that is, otherwise, too easily and too quickly overlooked. It is about the wholeness of the realization of something. Inversion is a model for people to imagine potential conflicts and, using the example of the opposite, recognizing potential problem points in the first place, as well as to try out and tackle their solutions. Inversion is also a model that can be used to achieve maturity and foresight on the part of the conflicting parties. The perspective of the other person becomes relevant to oneself. The feeling for oneself expands into compassion for the other person, who is put into a fragile state by the inversions, just like oneself. Since a conflict is never absolutely isolated from the events of life, as a whole, it can also be recognized and experienced as such by other people in the broadest sense. In other words, just as my fellow human being behaves in a conflict, I could now recognize myself in this other human by means of inversions; the same applies to the landscape as a shared living space. Even a self-oriented feeling such as a sense of self only exists because I know myself in contrast to the other. However, this also means that if my sense of self includes a sense of sharing, I am connected to the other on this basis in such a way that I can share in their suffering or joy. The feeling enables undisguised participation in the other and this shared world (cf. Emrich und Roes 2011). You recognize yourself in the other person and are now able to understand their actions to a certain extent. Inversion strengthens empathy. It makes it possible to empathize with and in the other and, in this connection with the other, to be jointly focused on something that affects both in conflict. "By empathizing with the other, I perceive myself as an objective self" (Theunissen 1977, p. 85), according to Husserl. Furthermore, "I not only empathize with his [of the other] thinking, feeling, doing, but must follow him in it, his motives become my quasi-motives", i.e., the empathizing of I must "participate" with them (Theunissen 1977, p. 86).

From the perspective of *ideological* criticism, the inversion shows that the world is never just 'one'. The world is multifaceted. Since a conflict is rarely just a single conflict, but usually has varied impact on other people, it is all the more important to gain a variety of insights into the conflict and to develop the ability to create a variety of possible solutions. In the age of "apparent diversity" (Bauer 2018, p. 9), inversion is a practice in diversity.

From the perspective of a worldview critique regarding the *consequences and unintended side effects*, it can be stated: By renouncing inversions, one would deprive oneself of the possibility, as shown here using the example of landscape, of demonstrating the enormous diversity of this area of life that is constitutive for the individual. Neglecting diversity leads to one-sidedness, which brings with it the danger of dogmatism. A way out is made possible by the emotions affected by inversion as a perception of possibility in the sense of an open mind. Robert Musil writes: "But if there is a sense of reality, and no one will doubt that it has a right to exist, then there must also be something that can be called a sense of possibility. Those who possess it, for example, do not say: Here this or that has happened, will happen, must happen; but they invent. Here could, should or must happen; and if you explain to him that something is the way it is, then he thinks: Well, it could probably be different.

Thus, the sense of possibility could be defined as the ability to think everything could equally well occur and not to contemplate what is more important than that which is not. One can see that the consequences of such a creative disposition can be remarkable; and, unfortunately, they often make what people admire seem wrong and what they forbid seem permissible, or even perceive both indifferently. Such men of possibility live, as they say, in a finer web of mist, imaginations, reveries and subjunctives" (Musil 2014, p. 16). Inversions are exercises in the sense of possibility. Thinking within the horizon of the possible means seeing more, not less. Those whose thoughts are motivated by the 'possible', do not think of something as completely fantastical, but see the same objects of experience in a larger context. They envision more because they do not stop at the boundaries of what is factually necessary but see the factually necessary as having open boundaries that invite them to think further. This further thinking is based on the conviction that the factual reality cannot force one absolutely, but that one has the ability to go beyond the constraint and thereby prove oneself against the constraint of the conflict. Learning how to overcome conflict would be the invention of possibilities.

From the perspective of *life-related pragmatic* criticism, the following should be noted: Emotion is so important because it indicates the most direct way in which an individual is affected by a conflict. Moreover, in the capacity of emotions, we are inextricably linked to the environment and to 'the other' before any reflection and thus before any potentially one-sided analysis of the conflict. In the feeling of immediate existence, the conflict situation manifests itself undisguised and in one fell swoop. Building on this, the aforementioned strategies of inversion can then be used to evoke counter-images to the current situation, which evoke feelings of foreboding, longing, etc., in the individual, motivating them to create new possibilities and develop these with regard to collective values (Berr et al. 2023; Kühne et al. 2023).

References

Appleton, J. (1975). *The experience of landscape.* London: Wiley.

Appleton, J. (1984). Prospects and Refuges Re-Visited. *Landscape Journal 3* (2), 91–103. https://doi.org/10.3368/lj.3.2.91

Bauer, T. (2018). *Die Vereindeutigung der Welt. Über den Verlust an Mehrdeutigkeit und Vielfalt* (3., erneut durchgesehene Auflage). Ditzingen: Reclam.

Berr, K., Lohmann, P. & Kühne, O. (2023). The Contributions of Philosophy and the Social Sciences to Landscape Conflict Research—A Critical Comparison. *Sustainability 15* (24), 16802. https://doi.org/10.3390/su152416802

Bourdieu, P. (1987 [1979]. *Die feinen Unterschiede. Kritik der gesellschaftlichen Urteilskraft* (Suhrkamp Taschenbuch Wissenschaft, Bd. 658). Frankfurt (Main): Suhrkamp.

Derrida, J. (2003). *Eine gewisse unmögliche Möglichkeit, vom Ereignis zu sprechen.* Berlin: Merve-Verlag.

Emrich, H. M. & Roes, M. (2011). *Engel und Avatare: Ein Dialog über reale und virtuelle Welten.* Berlin: Matthes & Seitz.

Frank, M. (2002). *Selbstgefühl. Eine historisch-systematische Erkundung.* Frankfurt (Main): Suhrkamp.

Holmén, H. (1995). What's new and What's Regional in the 'New Regional Geography'? *Geografiska Annaler: Series B, Human Geography 77* (1), 47–63. https://doi.org/10.1080/04353684.1995.11879680

Kaplan, R. & Kaplan, S. (1989). *The experience of nature. A psychological perspective.* Cambridge: Cambridge University Press.

Kneer, G. (2009). Jenseits von Realismus und Antirealismus. Eine Verteidigung des Sozialkonstruktivismus gegenüber seinen postkonstruktivistischen Kritikern. *Zeitschrift für Soziologie 38* (1), 5–25. https://doi.org/10.1515/zfsoz-2009-0101

Knoblauch, H. (2016). Über die kommunikative Konstruktion der Wirklichkeit. In G. B. Christmann (Hrsg.), *Zur kommunikativen Konstruktion von Räumen. Theoretische Konzepte und empirische Analysen* (S. 29–53). Wiesbaden: Springer VS.

Krohn, W. (2000). Wissenschaftssoziologie. Zwischen Modernisierungstheorie und Sozialkonstruktivismus auf schwankendem epistemischem Boden. *Soziologische Revue 23* (Supplement). https://doi.org/10.1524/srsr.2000.23.sonderheft5.314

Kühne, O., Berr, K. & Lohmann, P. (2023). *Landschaft zwischen Philosophie und Sozialwissenschaften. Eine Kritik.* Wiesbaden: Springer Fachmedien; Springer VS.

Lewin, K. (1927). Untersuchungen zur Handlungs- und Affektpsychologie. *Psychologische Forschung 9* (1), 1–85. https://doi.org/10.1007/BF02409755

Lohmann, P. (2006). Gefühl und Selbstaufklärung: Genetische und faktische Ausprägungen der Gefühlstheorie J.G. Fichtes. *Archiv für Begriffsgeschichte* (48), 111–135.

Mieg, H. A. (2001). Verantwortungsethik. In J. Ritter, K. Gründer & G. Gabriel (Hrsg.), *Historisches Wörterbuch der Philosophie. Band 11: U-V* (S. 575–576). Basel: Schwabe.

Musil, R. (2014). Der Mann ohne Eigenschaften. Erstes und zweites Buch, Bd. 1. In A. Frisé (Hrsg.), Hamburg: Rowohlt-Verlag.

Ng, E. (2020). No Grand Pronouncements Here…: Reflections on Cancel Culture and Digital Media Participation. *Television & New Media 21* (6), 621–627. https://doi.org/10.1177/1527476420918828

Oesterreich, P. (2011). *Spielarten der Selbsterfindung: Die Kunst des romantischen Philosophierens bei Fichte, F. Schlegel und Schelling.* Berlin, New York: de Gruyter.

Orians, G. H. (1980). Habitat Selection: General Theory and Applications to Human Behavior. In J. S. Lockard (Hrsg.), *The Evolution of Human Social Behavior* (S. 3–25). New York: Elsevier.

Orians, G. H. (1986). An Ecological and Evolutionary Approach to Landscape Aesthetics. In E. C. Penning-Rowsell & D. Lowenthal (Hrsg.), *Landscape Meanings and Values* (S. 3–25). London: Allen & Unwin.

Rorty, R. (1997). *Achieving our country. Leftist thought in twentieth-century America* (The William E. Massey, Sr. lectures in the history of American civilization, Bd. 1997). Cambridge, Mass.: Harvard University Press.

Schwarz, N. & Krex, A. (2022). “Riecht es nach Fisch, sind die Menschen misstrauischer”. Interview mit Norbert Schwarz in Zeit online vom 22. April 2022. https://www.zeit.de/entdecken/2022-04/intuition-psychologie-misstrauen-sozialpsychologe-norbert-schwarz/komplettansicht. Zugegriffen: 29. Juni 2024.

Stalfort, J. (2013). *Die Erfindung der Gefühle. Eine Studie über den historischen Wandel menschlicher Emotionalität (1750–1850).* Bielefeld: transcript.

Stegemann, B. (2018). *Die Moralfalle. Für eine Befreiung linker Politik.* Berlin: Matthes & Seitz.

Theunissen, M. (1977). *Der Andere: Studien zur Sozialontologie der Gegenwart.* Berlin, New York: de Gruyter.

Chapter 6
Valuing Criticisms: Extending and Specifying Inverse Landscapes and Their Representations

Previously in Chapter 5 the concept of inverse landscapes was subjected to critical reflection, as based on the data presented in Chapter 4 which serves to visualize and illustrate landscape contingency as well as its translatability into concrete representations. This also applies to the connection back to the (carto)graphic creation of inverse landscapes in relation to the (self-)ironic attitude that is particularly relevant in neopragmatism. Last, but not least, the integration of emotion into the concept of inverse landscapes has been subjected to the scrutiny of critique. In the following, we will deal with the formulated criticism, follow it, reject it, use it as a starting point for the further development of the concept of inverse landscapes and their (carto)graphic representation, and also with neopragmatist landscape theory as a whole. The underlying broader central questions also include: To what extent are pragmatism and linguistic theory a useful basis for landscape science? Are the categories of critique useful? An answer to these questions will be provided in the conclusion (Chapter 7) in an unequivocal manner.

6.1 Taking up Criticism 1: Prerequisites for a Neopragmatic Landscape Science

If the pragmatist standard of usefulness is applied, *internal criticism* as conveyed in Sect. 5.1 can be understood effectually. At the same time, it is also necessary to note that the high degree of complexity that arises from a neopragmatist approach to landscape is a consequence of the complexity of the subject matter; as this book makes clear, landscape has tangible, individual consciousness-related, socio-cultural, and intersubjective connections. No single theory, method, dataset, disciplinary perspective, or method of representation can make this complexity manageable. The extensive triangulations help to minimize the danger of falling into a causalist or representationalist trap and preserve plus expand contingency awareness instead. Triangulation

O. Kühne et al., *Neopragmatism - Inverse Landscape - (Carto)graphic Representation*, Neopragmatic Horizontal Geographies 2, https://doi.org/10.1007/978-3-031-89402-2_6

of researcher perspectives is recommended to reduce the burden of knowledge and evaluating theories, methods, et cetera.

As concerns the questions outlined in Sect. 5.1, the statement that neopragmatism is relativistic, resulting from the perspective of essentialist and positivist science, can be rejected. It is not relativist, but pan-relationalist, i.e., it is tied back to proven vocabularies that are further developed and in some cases redescribed in order to generate new useful solutions for changing societies (Rorty 2023). In doing so, the idea of an 'objective truth' is abandoned for well-founded reasons. First of all, man is not in a position to grasp 'the world out there' as it is sensed and certainly not to depict it representationally; man is causally, but not representationally, linked to this world (Rorty 1980, 2023). If we abandon representationalist ideas, we gain the opportunity to develop greater perceptiveness for different vocabularies. In addition, the focus on suitability reduces the danger of intense and brutal conflicts, because a pan-relationalist view is always aware of the contextual dependency of knowledge. In this respect, the neopragmatist understanding of science is not that of a neutral 'observation of the world'. Rather, science is a tool for generating useful knowledge (whereby this is typically aimed at orientational knowledge, rather than dispositive knowledge). Bringing theory back to and into practice reduces the danger of self-referential problem formation (philosophy for philosophers). The pragmatist roots of neopragmatism thereby also prevent the danger to which purely constructivist theories (including linguistic theories) expose themselves—that of (pure) self-referentiality.

The *metatheoretical* criticism in Sect. 5.1 refers to the tendency to reduce complexity and to contain contingency through neopragmatism (here in spatial and landscape research). This danger can be described as inherent in any theory formation. This is because every theory formation involves abstraction. This creates the danger of reducing complexity excessively, but also of replacing contingency with a binary coding of an 'either/or'. In neopragmatism, these dangers are mitigated on the one hand via theory triangulation, but also by sensitivity to hybridity. Accordingly, in neopragmatist research, results and representations must be examined to determine whether they are not too closely tied to representationalist thinking, unduly reduce complexity, and diminish contingency. Moreover, the aim of neopragmatist research is to produce such redescriptions that can be woven into further redescriptions. In this respect, they can provide a basis for being woven into speech descriptions that are even more strongly oriented towards familiarizing and generating contingency. The categorical separation of public and private can also be used as an example of such a redescription. This dichotomous interpretation of the world forms the foundation (conceptual) that can be differentiated today because of public and private hybridization. Today, private self-creation and public solidarity can be thought of as a polarity rather than a dichotomy. The criticism that neopragmatism is limited to the function of a meta-stage regulation mechanism for dealing with vocabularies is partly justified. Neopragmatism functions as a metatheory in landscape research based on it to date, but the first steps have also been taken towards a pan-relationalist landscape theory. For example, (carto)graphic representations of spaces interpreted as Landscape 1 provide the potential to use (carto)graphic representations as perceptual guides and normative standards for intervening in material spaces. The concept of inverse landscapes (including their (carto)graphic representations) is also a contribution to the development of a pan-relationalist landscape theory—towards the development of which this book is certainly intended to contribute.

Some gladly accept the criticism voiced from *an ideological perspective* that neopragmatism, with its focus on practical benefits, is not suitable for the construction of 'utopias', because—beyond his meta-utopia of an inclusive society—his falsificationist thinking demonstrates an ideological incommensurability with worldviews involving an affirmative reference to utopianism and a closed society. Accordingly, such individuals also accept criticisms of a liberal worldview's basic position of expanding the 'we' and avoiding pain; after all, these individuals are oriented towards a society that is a liberal democracy.

Following the ideological criticism, the statement that utilitarian thinking and falsificationism are not suitable as an integrative idea for the future of a society can also be affirmed regarding the criticism of *consequences and unintended side effects*. Neopragmatism does not formulate a 'grand narrative' (Lyotard 1979). On the contrary, if its falsificationism is related to utilitarian thinking, it can even be extended: Not everything need be immediately useful for solving practical problems. Instead, two aspects can be extended: First, the loss of usefulness can be less than in the case of world interpretations of grand narratives (put differently, its harm must be less than that which results from the pursuit of an inappropriate 'grand narrative', and the harm that has resulted from grand narratives is enormous, as is clear not least from the attempts to impose national socialism or communism). Second, the benefits can lie at the metalevel, for example, in the removal of obstacles that hinder the generation of useful solutions to practical problems, i.e., the removal of dysfunctions. In our context, this means, for example, removing the limitations of critical cartography in a neopragmatist cartography.

A similar argument can be made regarding the neopragmatist focus on procedural justice, which does not contemplate the unequal distribution of life chances. However, procedural justice creates more life chances than it loses. Nonetheless, more life chances can be created through the utilization of life chances if the maximization of life chances is integrated into the conception of equitable procedures. The concept of life chances can be used—more generally—to extend and instantiate the neopragmatist focus on contingency. An extension is achieved by expanding contingency from World 2 and World 3, i.e., the 'world' of ideas, in the direction of social relationships, and an embodiment by using latent options thus expanded.

Neopragmatism must constitutively take *life-related pragmatic* criticism seriously; after all, it is aware of its roots in the everyday world and is, with its goal of finding useful solutions to concrete problems, ultimately oriented towards the everyday world, as well. It is precisely in order to develop useful solutions that 'intellectual shortcuts' (Korf 2022) remain as only the long and arduous path of dealing with complexity through a variety of theories, methods, data, etc. Since absolute certainty is impossible, only triangulation remains for achieving results that do justice to the complexity of the challenges. From this perspective, the task of science is not to produce 'certain knowledge about an objectively given world', but to expand the scope of contingency and thus also to contribute to increasing the openness of society. The idea of scientists as 'lifeless men' (Sloterdijk 1987), who hover over things and depict these things objectively, is replaced by the recognition of the contextualization of things—also with researchers. Accordingly, it is not so much the 'desire for truth' that becomes the central qualification, but rather the willingness to think from multiple perspectives. The fundamental centering of 'we' as an approach strengthens the attitude of collective uncertainty to enhance further progress. Instead of self-centeredness, individuality (Reckwitz 2017, 2019), and an ego-protective "relationship of aggression to the world" (Rosa 2019, p. 41) being promoted, an accepting attitude that growth is only possible in the context of respectful encounters and communication with others, which does not fight over stereotypical positions and being right (cf. Popper 1996, p. 161), preferring alternatively to ensure "that more and more diverse people fall under our use of the word 'we'" (Rorty 2023, p. 55).

6.2 Taking up Criticism 2: Neopragmatist Understandings of Landscape

On the subject of the information presented in Sect. 5.2, the following can be stated: The world in general, landscape in particular, exhibits a high degree of complexity. The structuring into Landscape 1, Landscape 2, and Landscape 3, as implemented by the differentiated Mode A, Mode B, and Mode C allocations, results in the construction of a categorizing approach that facilitates the handling of contingency and complexity without striving for a decomplexification or destruction of contingency.

As a result of the high degree of complexity, neopragmatist landscape researchers are faced with the challenge of clarifying the usefulness of the approach. This is done against the backdrop of the topic, research question, and target audience. However, it is essential that the neopragmatist approach to landscape research generates knowledge that offers guidance rather than knowledge that is available, which is geared towards dealing with the complexity and contingency of landscape. This also highlights the connection between usefulness and contingency: both its awareness and creation.

The information in Sect. 5.2 was brought forward (as an example) in relation to evolutionary landscape theories and critical landscape theory. Evolutionary landscape theories are problematic for neopragmatist landscape theory for two reasons: First, since they speculatively assume that (present-day) landscape preferences emerged as configurations of Space 1 experienced in prehistoric times, they are immunized against falsification (since interviewing prehistoric humans is impossible according to the current state of science). Second, by assuming a biology inscription, they tend towards essentialism (whereby essentialist theories generally tend to immunize themselves against falsification efforts, since they aim at the 'true nature' of a 'thing' and once the truth is recognized, falsification efforts are obsolete). In this respect, evolutionary theories lie outside the (broadly defined) boundaries of what can be included in neopragmatist theory triangulation. This does not mean that they are not useful for neopragmatist landscape research on the one hand, they can be used as objects of research, on the other hand, they serve to clarify where the boundaries of the vocabulary of neopragmatist landscape research are drawn. Concerning the criticism from the perspective of critical landscape research of the insufficient centrality of power relationships, two central aspects of neopragmatist landscape research can be pointed out: First, the dependence of the choice of theory on topic and question(s), and second, usefulness. Thus, if the topic and question(s) and/or the expected usefulness of results make the inclusion of the topic of power appear reasonable, questions of power will also be addressed. In this respect, critical landscape theories can also be included in the research, or the connection between power and landscape can also be included in the theory triangulation beyond critical theories (after all, from a neopragmatist perspective, these do not have an exclusive claim to the investigation of the device that is 'power').

From a neopragmatist perspective, however, it seems reductionist to want to restrict all landscape relationships by power relationships. In this kind of science, the assumption that all relationships are determined by power has taken on a function equivalent to that of the all-pervading Satan (Rorty 1997).

The *metatheoretical* criticism of the quasi-ontological categorization by means of the theory of the three landscapes is justified. Both points require a conceptual extension. At this point, we will deal primarily with the first criticism and expand the theory of the three landscapes to include an intersubjective component. However, to add an intersubjective component to the neopragmatist landscape theory, we will only make a few basic considerations that this extension would demonstrate a dimension similar to that required for the development of the neopragmatist metatheoretical approach to landscape research. Let us begin with the extension of the theory of the three landscapes.

The co-evolution of landscape does not take place solely between Landscape 3 and Landscape 2 or in the confrontation of Landscape 2 in the categorization, interpretation, and evaluation of a Space 1 as Landscape 1 (with recourse to Landscape 3), but intersubjectively through different Landscapes 2 with recourse to different contents of multiple Landscapes 3 (whereby the different modes are actualized in the process). The starting point here is the existence of different Landscapes 2, which arise on the one hand from the fact that not all contents of Landscape 3 are socialized, but only parts of it (as with the remarks on Mode A, Mode B, Mode C, as well as the linguistic/cultural differences comprising the production of Landscape 3 have made clear, see Sect. 2.2). Different people also have different experiences with Spaces 1, also with different Spaces 1 interpreted as Landscapes 1.

The result is that many different Landscapes 1 are synthesized into a singular Space 1 by different people (Fig. 6.1). These different landscape syntheses, in turn, are not incommensurable in principle; after all, we are dealing with a linguistic version that does not exhibit similarities involving objects and contextualizations (for example, individual tree leaves are hardly ever included in landscape syntheses, the track

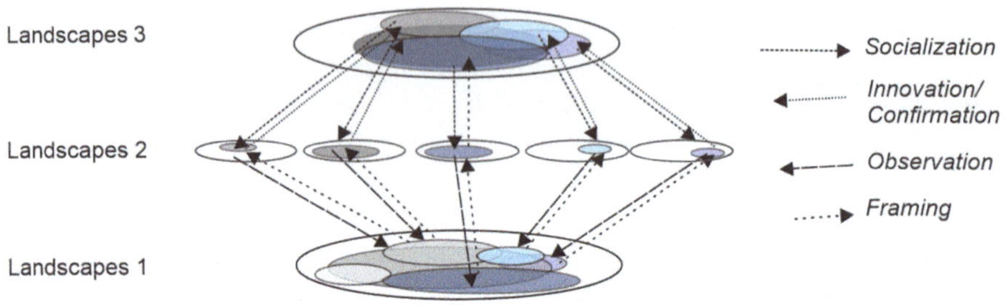

Fig. 6.1 The differentiation of 'Landscapes 2' (Figure based on: Kühne 2020)

diagram of an electric locomotive or a pile of empty wine bottles are described as a landscape; on the general commensurability of language: Davidson 2004).

The large number of different Landscapes 2 makes it possible—as shown in Fig. 6.2—that an inverse Landscape 2 can emerge not only through the individual's engagement with a Space 1 that is categorized, interpreted, and evaluated as Landscape 1 and ultimately lead to an expansion of the individual's understanding of landscape (or through engagement with Landscape 3), but (also) through intersubjective interaction. This intersubjective communication takes place through the use of World 1 as a medium, since direct communication between Worlds 2 (and thus also Landscapes 2) is not possible; after all, these form operationally closed units (Kneer und Nassehi 1997; Luhmann 1984). This communication process takes place through the translation of the contents of the subjective consciousness of the imaginary language within ALTER into spoken language, which is transmitted to EGO by means of the vibration of air molecules (alternatively: letters on paper, electronic impulses, etc.), who transmits these by means of hearing (alternatively: seeing) in World 1 located outside its own body to the internal World 1 as nerve impulses and translates these into the consciousness of EGO (the process can continue reciprocally). Provided that ALTER and EGO are able to speak a common language (and the conditions of external World 1 allow this, for example by the absence of a high noise level or other distractions), EGO and ALTER should be able to intersubjectively identify similarities and differences with regard to the interpretation, evaluation, and categorization of landscape (generally speaking: Davidson 2004).

In a process of exchange aimed at clarifying or expanding the contingency of landscape, ALTER makes the offer of an inverse landscape, which EGO accepts and expands its Landscape 2 (Fig. 6.3).

However, such an effect of inverse landscape is not mandatory (Fig. 6.4). Here, for example, the offer of an inverse landscape reaches a person whose Landscape 2 is dichotomously demarcated from Not-Landscape 2 and there is no willingness to shift this boundary in the direction of Not-landscape 2. A prominent example within landscape research is the so-called 'Körner–Eisel–Trepl debate', which revolved around Martin Prominski's dissertation *Landscape Design* (*Landschaft Entwerfen*, Prominski 2004) and the concept of the so-called 'Landscape Three' (Eisel und Körner 2009; Jackson 1984) as well as the question of whether contingent Landscapes

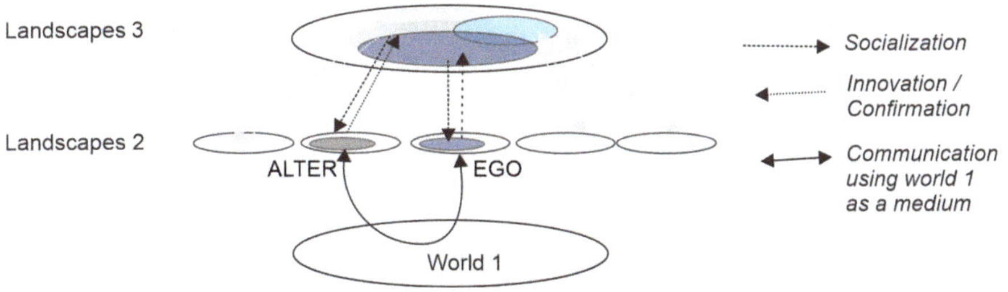

Fig. 6.2 The intersubjective exchange of individual landscape perceptions (Landscapes 2), here between two people. Other individual understandings of landscape have no particular significance in this context and are shown in a correspondingly transparent manner (own illustration)

2, which do not correspond to established notions of landscape (such as urban spaces or old industrial landscapes), but rather as Not- Landscapes 2, can also be included in the understanding of Landscapes 2.

The more fundamental criticism of the use of the quasi-ontological categorization by means of the (also now extended) theory of the three landscapes instead of a pan-relationalist conception of a neopragmatist landscape theory cannot be extended by a simple extension of the existing approach alone but requires broad conceptual considerations. These are—from today's point of view—quite reasonable for making the possibilities of neopragmatism available for landscape research (more broadly also: spatial and social research). A pan-relationalist neopragmatist landscape research would not only complement the metatheoretical perspective developed to date but would also expand the fundamentally conceived vocabulary theory of neopragmatist landscape research to include material components. This can be linked to fundamental considerations of assemblage and actor-network theory (Allen 2011; Hilbig 2024; Jahncke und Sprißler 2024; Kühne 2019; Winsky 2024) whereby these should be expanded to include the utility orientation, which represents the pragmatist legacy of neopragmatism. The roots in pragmatism also make the use of the (now expanded) theory of the three landscapes appear opportune, even if they cannot be integrated seamlessly into a pan- relationalist approach: It has proven its worth in numerous studies on landscape (in particular) and regional geography (in general) (for example in: Berr et al. 2022; Berr et al. 2023; Berr und Kühne 2023, 2024; Koegst 2022; Kühne et al. 2024; Kühne, Berr, Lohmann 2023b; Kühne und Koegst 2023).

Against the background of language organized in a representational manner (Rorty 1997 [1989], 1998) and the dominance of binary conceptual distinctions, the theory of the three landscapes can be understood as an intermediate step that is connectable to other perspectives (and thus potentially useful), but which will also (probably) become obsolete with the development of pan-relationalist neopragmatist landscape research.

The procedure described in Sect. 5.2 was followed in part with regard to the dominant significance of Landscape 2 by adding the intersubjective to the theory of the three landscapes in the previous section. Nevertheless, the individual remains intrinsic (which is not surprising in a liberal theory). The criticism that the material

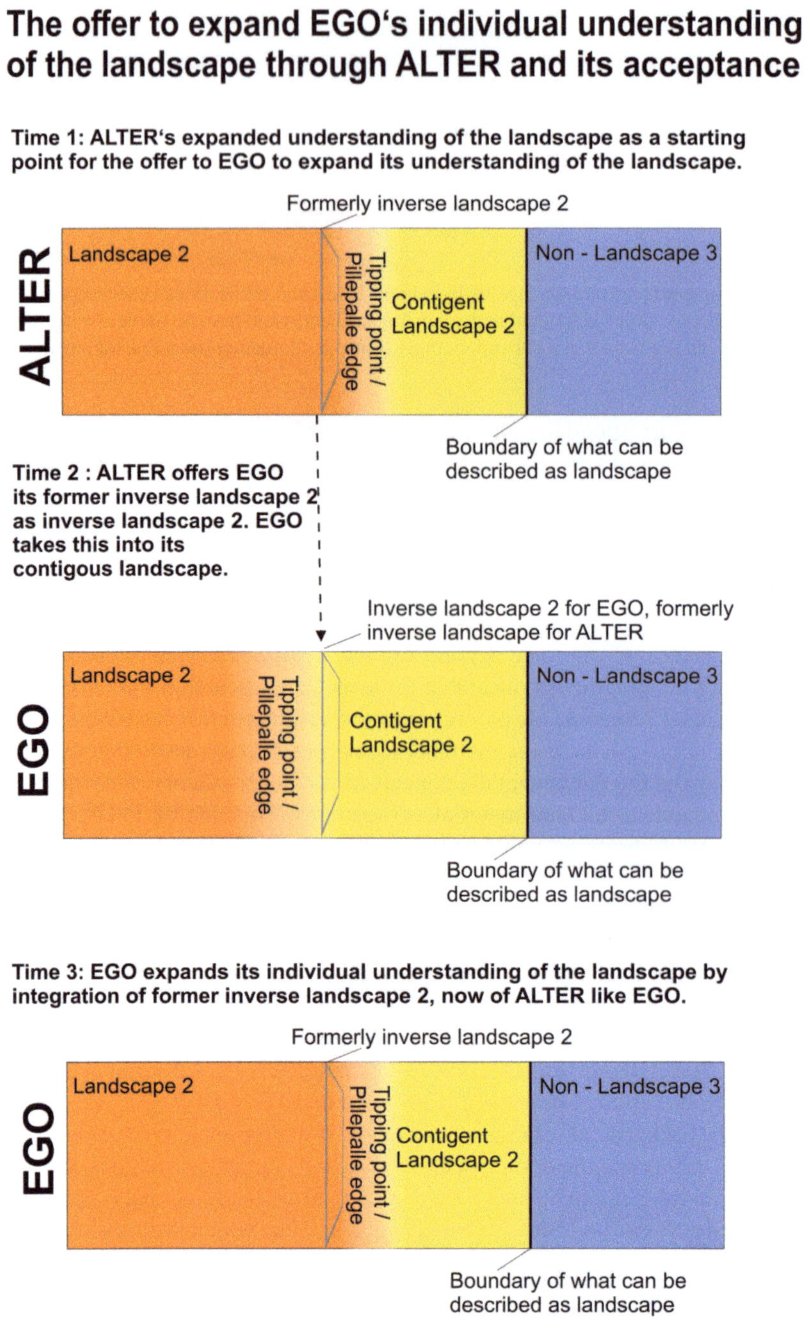

Fig. 6.3 The expansion of the 'Landscape 2' of EGO as a result of the offer of an inverse 'Landscape 2' by ALTER (own illustration)

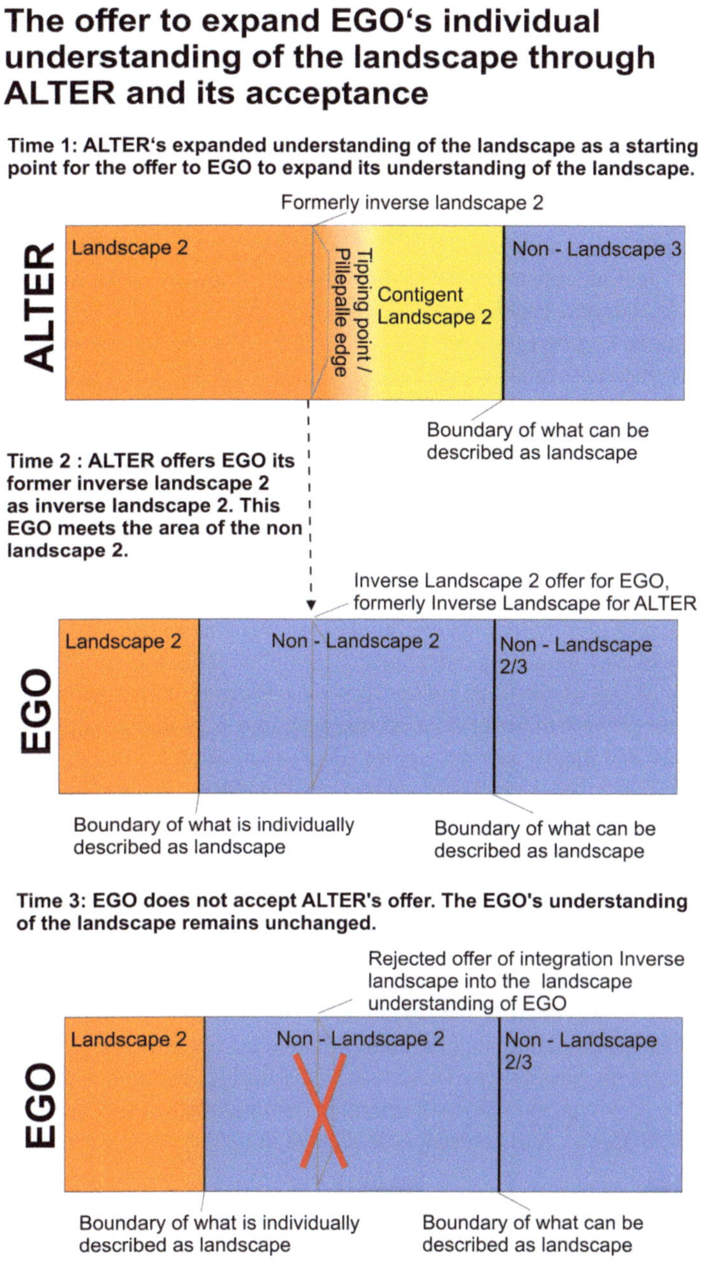

Fig. 6.4 The rejection of the extension of 'Landscape 2' by EGO as a result of the offer of an inverse 'Landscape 2' by ALTER (own illustration)

has no inherent meaning must be accepted. This is the consequence of the fact that the 'world out there' has a causal, but not a representational meaning in relation to the human being.

As regards the information provided in Sect. 5.2, it can be stated that there are *consequences and unintended side effects*: Neopragmatism, in general, and neopragmatist landscape theory in particular, is tied to efforts to deal constructively with contingency and complexity, not just to recognize and raise their potentials. It lacks both the possibilities and the will to force people to do so. It can only point to the effect of dealing constructively with contingency and complexity in a way that increases life chances. Problems that require (or appear to require) a clear solution should therefore first be examined to determine to what extent contingent solutions might not be more useful. This applies, for example, in the context of dealing with Space 1: whereas the aim of modern planning was to manifest monovalent uses (either commercial or residential) in Space 1, planning that is oriented towards the potential of contingency strives to manifest polyvalent uses (both commercial and residential), for example, to be able to compensate for the consequences of economic structural change with less effort (cf. for example Ellin 1999; Fontaine 2017; Kühne 2006, 2016; Kühne, Jenal, Koegst 2020b).

The life-related pragmatic critique must take seriously an approach that sees itself in the tradition of pragmatism. However, it is not possible to react to an increasingly complex world with excessive decomplexification, which could not do justice to the contingency of the world in any way. However, neopragmatist landscape research must not only present the results of its research in a way that is appropriate to the subject matter and theory, but also appropriately in relation to the target audience. If the target audience has a primarily subject specific Mode C approach, the demand for general comprehensibility is lower. If it is primarily intended to reach people with a Mode B approach, the requirement is higher; this also applies to interdisciplinary comprehensibility, since—as shown—specialist vocabularies can be essentially traced back to everyday language and can thus be translated into each other (Putnam 1997). The cognitive tendency towards decomplexification can be countered by a culture of conveying facts and threads of subject specific discussion incorporating a Mode C approach in the direction of Mode B that is adapted to everyday language. The fundamental neopragmatic orientation towards the 'extension of the we' also creates the possibility of resorting to (figurative) communicative means that transcend the previous boundaries of scientific communication and encourage people with a Mode B approach to landscapes to engage in further discussion about (future) landscapes.

6.3 Receiving Criticism 3: The Concept of Inverse Landscapes

As regards the formulated internal *criticism*, the comment regarding the opposition to the manifestation of inverse landscape in relation to the contingency idea of neopragmatism can be followed in principle. However, this following relates in a distinct way to the dimension of the linguistic theoretical foundations of neopragmatism. Its pragmatist roots with their reference to usefulness regarding the solution of actual problems presuppose, however, that recipients of ideas—notably here those conveyed in visual language—can also accept communication (here in the terminology of Luhmann 1984). In other words: the representation of inverse landscapes must be connectable for them so that they can integrate them into their communicative processes. In this context, their function lies in the implementation of the abstract. In this respect, the criticism of pictorial representation expressed from a more-than-representational perspective is also justified, although the pragmatic ability to operationalize must also be considered here. Again, as long as language (including signature language) is composed in a representationalist way, we can criticize its reductionism, yet we are still dependent on it as a tool of communication.

The criticism from the *contextualization* can be followed. First, inversion can be divided into historical, present, and future (utopian with different assumed degrees of commitment) relative to the respective points in time. Therefore, in relation to the current state, inverse landscapes are formed as historical, present, or future. Regarding the future generation of inverse landscapes, the openness of the future must be considered, which means that future inverse landscapes may not yet be conceivable today. This is where the problem, of existent utopian inverse landscapes, becomes clear: they entail the danger of preventing the creation of future inverse landscapes that are not yet visible today, since teleologically only one inverse landscape is understood as legitimate. Second, inverse landscapes exhibit a great variance of contingency, from narrowly guided landscape forecasts to speculative visions. Third, inverse landscapes have varying potential to have a socially functional effect (here following: Edler und Kühne 2022): They can have a contingency expanding effect, i.e., functional and meta-functional effect, and they can have a contingency restricting (dysfunctional) effect. However, they can also have no effect and thus remain afunctional. A special case in this context is when inverse landscapes restrict contingency to make contingency recognizable. In this case, they in turn expand contingency and have a meta-functional effect (neopragmatist ironic cartography, for example, can claim this for itself when it depicts spaces in a reduced manner to stimulate the question of which representations have been dispensed with).

The internal critique of inverse versus not-inverse hybrids leads on to the *metatheoretical* critique, as against the theoretical background it is simply obvious not to make a dichotomous distinction between landscape and inverse landscape, but to understand these as poles of a continuum, comparable to the 'Pillepallesaum' between landscape and contingent landscape. Against this background, landscape can also be partially inverse, i.e., only segments can emerge from landscape contingency.

Accordingly, being contingent and being not-contingent can be combined on all levels of landscape. Landscape 1 can contain aspects that do not correspond to conventional notions of landscape (here Landscape 3). In Landscape 2, the ability to connect to landscape interpretations, evaluations, and categorizations is increased if the inverse landscape does not take on a total form but is brought closer to World 2 in certain aspects. Involving Landscape 3, innovations can be incorporated into familiar understandings of landscape. The result is not a fundamental break, but a pastiche-like innovation. This in turn facilitates—in Richard Rorty's terminology—redescription.

Compared to metatheoretical criticism, *ideological* criticism is far less easy to follow: As has been emphasized on various occasions, neopragmatism and essentialism are two incommensurable positions. From the perspective of neopragmatism, no 'essence underlying the landscape' can be conceptualized, so the critique of essentialist positions remains without resonance in the context of the concept of inverse landscapes. Even the perspective of critical theory can only be followed in part: Neopragmatism formulates the meta-utopia of the expansion of 'we' and understands actual utopias, at best, as an expression of contingent thinking, at minimum, until the actual utopias make no claim to be universally binding.

The criticism voiced about the *consequences and unintended side effects* should not be dismissed. On the contrary, it must be taken seriously: Inverse landscapes conceived as exclusivistic, which lead not to a gain but to a loss of contingency, can neither be ruled out nor are they unlikely. Their manifestation in material spaces interpreted as landscapes is also within the realm of possibility, even if this involves the misuse of the concept of inverse landscapes to radically reduce contingency. Since this criticism is justified, but the danger can be mitigated by irony, although it can never be ruled out completely, all that remains is to remain vigilant, especially with regard to the unintended side effects, and to repeatedly point out the contingency that is constitutive of the concept of inverse landscapes.

The *life-related pragmatist* critique should also be taken seriously, as it addresses a fundamental and not easily synthesized challenge to the different thrusts of the basic positions of philosophy of language and pragmatism that form neopragmatism. While the former clearly differs from the everyday 'normal attitude' (Berger und Luckmann 1966) of 'the world is real' and can also be represented (linguistically), i.e., it deals with the complexity of the linguistic creation of the world, pragmatism expects the solution of existent problems and thereby also—as already emphasized—a communication of scientific considerations that are intersubjectively connectable beyond one's own narrow specialist vocabulary. From the perspective of the considerations on the social function of deviance outlined above, the inverse landscape can stand out strongly from the norms or expectations of society, as an entirety. This could, for example, describe a community that practices alternative values and lifestyles and therefore 'takes for a ride' the interpretations of society as a group ('mainstream'). This then contributes to the ability to question and potentially emancipate oneself from common stereotypes and thus has a meta-functional effect. If the life-related pragmatic contingency framework is thereby expanded, this means that at a later point in time, representations that now lie outside the landscape contingency can be embedded in the 'Pillepallesaum' and, hence, contribute to the expansion of the framework of contingent landscapes. This may still seem unsatisfactory from the point of view of current life-related pragmatic criticism, but it can also be understood as a regulation of the hardly resolvable different thrust of philosophy of language and pragmatism.

6.4 Taking up Criticism 4: Inverse Landscapes as Contingency—Potentials/Challenges

The *internal criticism* expressed specifies the fundamental dilemma that arises from the synthesis of neopragmatism, conceptually drawing on linguistic theory and pragmatism. The challenge of being committed to increasing contingency, on the one hand, and pragmatic connectivity on the other, has so far been solved by using simple signature language (symbology), reducing what is depicted to what is thematically and interpretatively necessary, using ironic elements, and visualized by borrowing from the ligne claire style. Other forms of representation are possible (and necessary). They should be developed and assessed to develop additional opportunities for the contingent handling of different types of visualization (also in combination with other sensory approaches).

The criticism from the *contextualization* can be followed. These are reasonable next steps that can contribute to an operationalization of the concept of contingent—and derived from this—inverse landscapes and their representation. Based on the layer principle, known from the processing of geodata (e.g., with cartography software or GIS), and also in the context of augmented and virtual reality, an inverse landscape could be a digitally generated environment, in the future also more process-automated through AI, which is created by algorithmic or digitally creative processes and which could generate a distortion or inversion of reality.

This changing of reality can have several dimensions in terms of design. Landscapes can deviate from a real scenario in terms of content and, for example, cancel out scientific and anthropogenically influenced regularities in their representation. One example is conveyed in Fig. 6.5.

Depicted landscape objects can also differ in their appearance or effect, either individually or in the overall structure, especially through the graphic variable of color, which is often used in the graphic sciences to represent qualities. An example that brings these two representational dimensions together graphically—and using AI—is Fig. 6.6.

Such distortions can be heightened by the (additional) visualization of visually non-experiential landscape phenomena. For example, acoustic phenomena that are scientifically measurable (e.g., noise) or culturally connotated can be represented in depictions of distorted landscapes. A whispering forest, for example, may culturally approximate a fairytale world, but move further away from a real landscape scenario (Fig. 6.7).

Fig. 6.5 Inverse landscape representation with unrealistic conditions. Penguins do not live in a landscape reminiscent of Ireland, and an urban business district depicted in the background is not indicated in this landscape scenario (source: own creation with DALL-E (Prompts: A realistic 3D landscape of the Irish countryside with penguins and a metropolitan area in the background.))

Fig. 6.6 Inverse landscape representation with underground sun. Scientific realities (e.g., light reflection) and object properties (especially color) are distorted here (*Source*: Own creation with DALL-E (Prompts: with inverted, surreal colors a representation of an underground sun (e.g., sun in a cave) with inverted, surreal colors))

Fig. 6.7 Inverse landscape representation of a whispering forest. Scientific realities (root growth) and object properties (especially color) are distorted here; the representation of 'whispering' adds both a non-visual and a culturally connotated layer (source: own creation with DALL-E (prompts: A representation of a whispering forest representation with inverted, surreal colors))

In the case of (carto)graphic representations using artificial intelligence, which therefore have no (directly human) author, contingency exists not only in the form of representation and the process of human interaction, but also fundamentally in the emergence of a new authorship of (carto)graphic representations. Depending on the training data of the AI, this can lead to overall social or partially social (carto)graphic forms of expression. This, in turn, can give rise to an examination of contingency on a meta-functional level. Artificial intelligence in particular can be used to generate a large number of representations of inverse landscapes in a very short time, with the result that, even according to individually created prompts, a large number of representations of inverse landscapes—even only gradually altered ones—can be generated quickly, for example, to explore the Pillepalle space or to draw the boundaries between inverse landscapes and not-landscapes (see Figs. 6.3 and 6.4). The same applies to actions that can be undertaken in such landscapes and represented cartographically by pictograms (Fig. 6.8).

Metatheoretical criticism can also be accepted: The presentation in the form of comics and maps or their hybrids is certainly not optimal as regards the goal of pursuing the more-than-representational approach of neopragmatism. However, it does represent a step towards questioning the omnipresent representationalist approaches to the world by providing an unusual form of presenting results, especially if they are (self-)ironic. This in turn reveals the dilemma of neopragmatism, the alignment of a philosophy of language geared towards increasing complexity with the pragmatist claim to provide solutions to concrete problems. This also highlights the particular challenge of operationalizing neopragmatism for dealing with an object

Fig. 6.8 Pictograms for cartographic representations created with DALL-E, here with a tendency towards the absurd according to Mode B access, namely throwing burgers at satellites (source: own creation with DALL-E (Prompts: Pictograms showing people throwing burgers at satellites))

that—because of its complexity—can hardly be grasped textually alone, namely space in general and landscape with its selective synthetic approach in particular (cf: Dodge und Perkins 2008, 2015). Accordingly, it becomes clear that the development of the depiction of inverse landscapes based on the ligne claire style can be one of many ways of depicting inverse landscapes. It represents an approach to testing in which the aim is to exemplify the possibility of shaping the cascade of instantiating contingent spaces and landscapes via inverse landscapes through to their pictorial representation in such a way that contingency is preserved. Based on this explorative pictorial manifestation of inverse landscapes, it is now necessary to advance the further development of this embodiment, also by extending it to other forms and techniques of representation, resulting from the use of augmented and virtual reality, the use of artificial intelligence, plus the use of alternative analogue representations, for example, in the form of painting, fine art, and hybrids thereof, but also by people who have a more Mode A and Mode B approach to the subject of space in general and landscape in particular.

Addressing the *worldview* criticism expressed, it can again be stated regarding essentialism: Neopragmatism and essentialism are two incommensurable systems of world interpretation. Since neopragmatism considers the 'essence' of something to be just as witless as the search for it, it is difficult to understand essentialist criticism as beneficial for one's own theory development. The critique is different from the perspective of critical theory: the depiction of inverse landscapes also represents an act of exercising power, as it is visualized according to the ideas of the depicter, but the constructedness becomes clear by means of ironic elements, and the claim to exclusive interpretative power is thus undermined. The concrete and thus seemingly determined representation is stripped of its unambiguousness by means of irony. In the context of neopragmatist landscape research—i.e., also the (carto)graphic translation of inverse landscapes—a further strategy for dealing with power- boundness is the precise justification of the steps leading to the presented result. Although this does not eliminate power, it does make the decisions that can be traced back to it explicit—and therefore just as explicitly questionable. For its part, the critique of power can be questioned—from a neopragmatist, and here also a pragmatist perspective. It ultimately produces an endless deconstruction of power processes, but deconstruction is also power-bound, as it claims a privileged (i.e., power-bound) insight into social contexts on a metalevel. This claim in turn can be deconstructed at the metalevel, which in turn presupposes privileged access at this level, which in turn is bound to power. The game of power criticism could be continued on all constructible metalevels, but this would hardly contribute to solving practical problems. This is where (neo)pragmatist critique comes in: Deconstruction of power processes is useful if it leads to the solution of social problems and does not represent a pure end, in itself (or, again, power-bound gain of distinction). In a neopragmatist interpretation, the diagnosis of the power-bound nature of social relationships does not just lead to decisions becoming explicit (and thus questioned). Rather, this diagnosis also focuses on the neopragmatist understanding of power: from a neopragmatist perspective, power is not a morally 'bad' aspect per se, but a mechanism that can be used productively in society and can also produce unproductive consequences and side effects. In this respect, it is important to empower the productive potentials and contain the destructive ones (Allen 2008). Hence, inverse experience means recognizing the self and the other by means of irony as a result of disturbance. Contingency is thus expanded under reactance when power is used as a motivator for further confrontation with contingency because of foreseeable or feared restrictions of freedom through the loss of alternatives (cf. "psychological reactance" according to Brehm et al. 1966) are associated with this. These confrontations with power create the opportunities to deal with power relationships, to accept them (voluntarily), or to work on their revision. This in turn leads to a problem of critical communication: The dedifferentiation of person and theory or the dedifferentiation of person and role (Flaßpöhler 2021; Grau 2017; Kostner 2019; Kühne et al. 2021; Kühne, Berr, Jenal 2023a) merge into a personal identity unit, whereby criticism of a theory is understood as an attack on personal identity, criticism of the performance of a role as moral devaluation. The result is a communication based on antipathies, which often starts from the point of view of an anticipatory stereotypical social and categorical generalization—instead of dealing with arguments. Such an approach, which can be characterized as vulgar ethics, is in turn opposed to neopragmatism with its norm of the process of integrative enlargement of the 'we' on the one hand and its focus on useful solutions on the other. This attitude goes hand in hand with communication practices that can be challenging, but at the same time are geared towards maintaining the integrity of the other person and respect for the other vocabulary. Appreciative criticism not only addresses the understanding of content (factual level), but also conveys a fundamental acceptance of the person. Rogers (2005) speaks of three basic characteristics for a growth promoting conversational framework on the sending side: empathy, congruence, and positive emotional appreciation. On the receiving side, Popper believes that the basis for a responsibility-based communication attitude is that growth only works if the interacting partners are challenged to contradict each other and are "more interested in learning than in being right" (Popper 1996, S. 169). The prerequisite for criticism is therefore that the content level is separated from the relationship level (cf. Schulz von Thun 2019). Since the understanding of development pursued here is based on the threefold contingency of language, self, and society plus—beyond inclusivism—has no utopian traits, it seems sensible not to fundamentally reject even a basic criticism of escapist approaches to the world (or turning away from it); after all, an escapist approach can also generate contingencies, as becomes clear not least in the 'game' (Huizinga 2019 [1938]; Kühne, Jenal, Edler 2020a; Longan 2008; Stintzing et al. 2020).

The criticism of the *consequences and unintended side effects*, that maps could be misunderstood as the design of a utopia, is equally justified, as the response to this has already been expressed on various occasions: A neopragmatist approach involves anticipating and reflecting on unintended side- effects. In this case, too, it is important to be aware of the danger of utopian interpretations. The (self-)ironic attitude towards oneself, language, and society should make the creation of concrete utopias more difficult, but it does not make them impossible. This becomes tangible in the (carto)graphic representation of inverse landscapes, as these are created in such a way that they (ironically) break with conventions of representation and the expectations of recipients.

The critique of the consequences and unintended side-effects hybridizes with the life-related *pragmatic* critique in the following. This was put forward on three levels: First, as regards the lack of ability to decipher the representation of inverse landscapes resulting from an unsuitable cultural chapter. This aspect especially relates to pragmatic orientation, i.e., the expected level of cultural capital of the target audience must always be incorporated when creating (carto)graphic representations. Neopragmatism can be understood as a philosophy that is concerned with the ability to deal with contingency and the productivity of contingency. In this respect, it is important to expand this ability, not to exclude it. Second, a desire for unambiguity (and thus the rejection of contingency) can be anchored in life practice. In this respect, it is the task of neopragmatist cartography (or other methods of presenting the results of research) to provide connectable representations that function as individually available latent options for dealing with the world. These may or may not become manifest options; this is for the individual to decide. Third, it cannot be ruled out that the pursuit of contingency is given any relevance at all in a life-related pragmatic sense, which means that dealing with contingency and its representation remains an expression of

Mode C vocabularies. Here, too, the approach of creating latent options that become individually available becomes clear. Whether these options are subjected to manifestation is then up to the person acting. On the one hand, it is therefore important to strengthen life chances by expanding latent options, especially individually available latent options. On the other hand, an approach to the world that is geared towards increasing contingency also means strengthening ethical ligatures, i.e., expanding the possibilities of questioning moral ligatures that are geared towards reducing contingency.

6.5 Taking up Criticism 5: Inverse Landscapes as Ironic (Self-)attitude—Potentials/Challenges

The *internal criticism*, as expressed, specifies in various aspects the fundamental dilemma of the double contingency of irony addresser and irony addressee or the contradiction between irony production and irony reception. This dilemma cannot be resolved by theoretical, possibly even a priori assumptions or suppositions, but requires comprehensive empirical research into perception and reception to better grasp and understand the constitutive prerequisites of the ability to produce irony and the forms of reception of ironic representations so that these can be adequately considered.

The subsequent criticism from the *contextualization* can be followed by specifying the answer to the internal criticism as a request to actually include the recipient in his contingent ability to receive ironic representations more strongly in the research design (Fig. 6.9).

Metatheoretical criticism can also be followed. When trust in irony arises from the insight that "there is nothing that we can rely on but ourselves and our fellow human beings" (Bernstein 2008, p. 22) and the 'self' of self-creation is not to be developed relative to identity or essentialistically, but always relationally, then the task of the liberal ironist extends beyond the realm of the private into that of the public and beyond "our relationships to other people" (Kretz 2023, p. 5). Rorty has linked this to the principle of solidarity, which derives from the basic moral attitude of compassion (Rorty 1997 [1989], 1999) with regard to clarification of the "expansion of our moral community" (2023, p. 55).

However, the constitutive conditions of the capacity for irony, which also lie in the socio-cultural and economic realms of the public sphere, must be distinguished from the objectives of the 'expansion of a moral community' and should not be conflated. The criticism of the personification of factual questions and the infiltration of the separation of person and role also belongs in this context of the distinction between the private and public spheres.

Fig. 6.9 The (partial) translation of an AI-generated image with a tendency towards unusual color choices into a 'ligne claire' representation underscores the intention to create a contingent interpretation of landscape by generating a graphic representation of an inverse landscape. In doing so, the context of typically available patterns of landscape approaches to the world is not consistently abandoned, in this respect the transition to not-landscape is rather exceeded in individual cases (source: own representation, based on the generation of DALL-E)

One objection to *ideological* criticism is that an insistence on strongly normative demands and a belief in immovable insights and knowledge as well as in corresponding 'no-alternative' concepts of action and decision-making—similar to essentialism—is incompatible with the basic convictions of neopragmatist irony. Positions that cannot or will not tolerate other positions alongside them (such as the god who does not tolerate other gods alongside him) ultimately withdraw from the competition between the most suitable concepts and positions at the present time and in relation to the problem.

One objection to the criticism of *consequences and unintended side-effects* is that a neopragmatist conception of irony is incompatible with conceptions that seek to overcome a supposed relativism by basing themselves on positions with an absolute claim to validity. The argument against a one-sided realist position is that a high metaphysical price would have to be paid for such a position.

The *life-related pragmatic* critique, on the other hand, can be followed insofar as scientific publications with corresponding ironic representations are not an end in themselves, but are innately 'stylized' from the lifeworld as a starting point and ultimately oriented towards the lifeworld as a target point. This results in the further consequence of respect for everyday world knowledge and the insight that Mode C knowledge is, with Putnam, founded in the language of the everyday world (Fig. 6.10).

Fig. 6.10 According to semiotic map theory, cartographic representations can be decoded not only syntactically, but also semantically and pragmatically: **Syntactically**, in recognizing that they are signs. **Semantically**, in recognizing that these signs can also be decoded in terms of their intended content, albeit within a certain framework of contingency. **Pragmatically**, in that signs, here an arrangement of signs in the form of a cartographic representation, can also be useful for practical actions. This pragmatic meaning becomes clear in the form of tourist maps, for example, which are intended to guide people to certain sights marked as such. This pragmatic dimension can also be simulated in contingent space, for example, as here, a (cartographic) representation of a long-distance hiking trail through our fictitious inversion, which connects three very geoecologically improbable spatial units along this path—especially in combination at a distance suitable for hiking: the **fluorescent dead sea**, the **sinking forest**, and the **lake of fire**. The images were created with the help of DALL-E and then edited in comic style using the graphics program CorelDraw SE. These were combined with a reduced representation of water surfaces, isohypses, and the course of the path. They are provided with pictograms, again created with DALL-E. Accordingly, this is a minimal representation, which would be necessary for the absolutely essential information for the hike. The inverse nature of the illustrations is reinforced by the choice of colors and the perspective arrangement of the comic-like illustrations and the cartographic representation. The choice of color and the comic-style depiction prevent the possibility of an aesthetic experience in the mode of sublimity inherent in the choice of motif by activating the mode of kitsch. The intersection with the (very reduced) cartographic representation (also with inadequacies in detail) also leaves little room for a feeling of sublimity. The aesthetic mode of sublimity suggests an 'essential' approach to the world and contradicts the aspiration of neopragmatism not to allow essentialist residues to take effect (Hagendorff 2011; Rorty 2023; Salaverría 2023) (*Source*: Illustration by Olaf Kühne based on the DALL-E illustrations, prompted by Dennis Edler)

With this aspect in mind, we can move on to the next section, which deals with the question of how emotion and compassion can be integrated into the concept of inverse landscapes within the framework of an interpersonal theory.

On the metatheoretical level, there is always the danger of wanting to identify an additional metalevel as a basis of justification for the first metalevel, and then similarly to this second metalevel, since it also requires justification (and is by no means absolutely valid under penalty of totalitarian thinking), and yet another one and so on, ad infinitum. This addresses a well-known problem in the philosophy of science, namely that the "problem of justification, which is unavoidable in any scientific theory formation, merely tends to iterate ad infinitum" (Gethmann 2023, p. 90) or transformed into a "technique of ethics [...] in the sense of a (higher-level) knowledge of the availability of orientation offers" (Hubig und Luckner 2013, p. 150). At the metalevel, proposed metatheoretical concepts or framework theories are non-absolute and contingent; they can therefore only be legitimized along the lines of purposes and goals that require justification. As a proposal, these must be fundamentally open to falsification according to Popper; at the object level, they can be accepted and implemented, or not. No metatheory can be justified absolutely.

6.6 Receiving Criticism 6: Unifying Emotion and Compassion in Inverse Landscape Concepts

The information in Sect. 5.6 can be constructively incorporated as follows: If the internal perspective of the individual in a conflict is not taken into account, this means that the conflict itself cannot be adequately understood. The individual himself is a composite of his life, his co-life with others, and his environment. Landscape is constitutive for it. Without landscape, there is no individual and it is directly as well as emotionally connected to the aforementioned components (Berr et al. 2023; Kühne, Berr, Lohmann 2023b). In this way, they experience themselves as unavoidable in their immediate lifeworld. If the original core of these entanglements is disregarded, it is impossible to know the full extent of a conflict for the individual and it is also impossible to help them sufficiently to decipher the meaning of the conflict itself, because they themselves know nothing of this entanglement, which for them is unprecedented and necessary. This is felt originally and in a certain respect also pre-reflexively. However, the entanglement itself must first be raised to consciousness. Only then is the individual able to take a meaningful stance on the conflict. Without this consideration, the individual himself would be incapable of acting and would merely be the object of the context of action that others develop in relation to a conflict and to which he must subordinate himself. Neither conflict nor solution would exist for this.

The criticism from the *contextualization*, as described in Sect. 5.6, can be engaged as follows: The inner motivations of the individual in conflict remain closed in Mode C. Seen from the outside, this also results in a certain helplessness in Mode C, because experts can only more or less guess what the conflict means for the individual. Precisely because they cannot go beyond assumptions—because they lack the internal perspective—more general and transferable solutions are offered. These solutions are merely external to the individual because the complexity of the conflict in its entirety, which results from the sum of the individual internal perspectives, is not known. Therefore, both the realization of the conflict and the solutions offered are never complete. They are merely indirect aids for the individual. The individual will hardly be motivated to participate in conflict resolution, not because he does not want to, but because he does not understand the conflict and its solutions in their full scope and cannot attribute them to himself and his lifeworld, nor can he know that he is understood by the experts of Mode C. Presumably, he will behave in a socially compliant manner in accordance with the guidelines resulting from Mode C largely out of fear of sanctions, but not out of actual insight. For this to happen, Mode A would have to be integrated into Mode B and these together into Mode C. In other words, the emotional state of the individual would have to be united with rational knowledge in a conflict on the part of the expert. This unification in the field of science, on the one hand, and in the field of popular communication of Mode C knowledge to Mode B, on the other, is also likely to confirm a number of Mode B stereotypes or make them appear questionable, which in turn reinforces or makes doubtful the individual's pre-refexively formative and action- guiding everyday experiences from Mode A. In this way, the individual could also assess his own biography regarding the conflict and thereby understand why he is or is not affected by a conflict and whether he should not change his established point of view, which is made up of his everyday knowledge and common-sense understanding. It is important to incorporate these preliminary stages of expert knowledge into Mode C of being in self-awareness and compassion in the immediate everyday life concerning the individual's willingness to accept inverse landscapes and, when constructing inverse landscapes, to focus from the outset on the biographical and personality-related prerequisites of the individual in Mode A and Mode B to avoid dysfunctional constructions.

On *metatheoretical* criticism: According to the above, we need a scientific model of landscape that integrates Mode A, Mode B, and Mode C, plus is differentiated into strict science (deductions, genesis of theories, a priori prerequisites of knowledge, and/or abstract concepts, theorems, etc.) and contrastingly the popular transfer of scientific knowledge into the factual, i.e., into the lifeworld. The direction here is not one-sided, because scientific knowledge has to take determinants of knowledge of Mode A and Mode B seriously and analyze them, i.e., in a certain respect also from everyday life as a "lifeworld a priori" (Mittelstraß 1974, p. 146) from which to start. The content of the knowledge of landscape stems from 'socially robust knowledge' (Nowotny 2000) of the everyday lifeworld and are available to everyone. They are only reflected in varying degrees in Mode A, Mode B, and Mode C. Ideally, the envisaged scientific model forms a circle: from life to science and from science back to life. Only a science applied in this way can adequately determine conflicts and formulate solutions that are appropriate to the individual and that can also be implemented by the individual. In this way, the individual is neither ignored nor overburdened. He is part of the whole and is personally affected by the conflict. Based on the definition of the role of philosophy in science from Mittelstraß, this envisaged model of science could be described as follows: 'Dispositional knowledge' in Mode A and Mode B would have to be given an 'orientating knowledge' (Mittelstraß 1995) with the knowledge in Mode C. Dispositional knowledge equates with competence in questions of concrete "knowledge of causes, effects and means" (Mittelstraß 1994, p. 37). It is thus on the side of "ability" (Mittelstraß 1994, p. 37). Orienting knowledge, on the other hand, is understood by Mittelstraß as "regulative knowledge of (justified) goals and purposes" (Mittelstraß 1994, p. 37). Orienting knowledge is thus on the obligational side of the "ought" (Mittelstraß 1994, p. 37). In this difference, both forms of knowledge necessarily complement each other. According to Mittelstraß, "both forms of knowledge belong [...] together. For without a regulative [...] knowledge [...] deficits in action arise, the ability that asserts itself in the knowledge available becomes disoriented" (Mittelstraß 1994, p. 37).

Therefore, "the rationality that modern developed societies need solve not only problems of ability, but also problems of 'ought'" (Mittelstraß 1994, p. 37).

With regard to *ideological* criticism, the following should be noted from our perspective: A counter- model to the model favored above would be the exclusive expert-like discourse in Mode C, for which the individual in his interpersonal relationships only plays a role as a figure of thought—e.g., in the context of sociological studies—but which pays no attention to the immediate life practice of these relationships in the reality of life (cf. Quast 2021). The individual as a living individual, who is capable of popular education in Mode A and Mode B, is a figure of thought abstracted from the reality of life for the discourse in Mode C and is determined by a theory. As a result, the individual is reduced to a function and not recognized as a feeling person, i.e., as a person directly affected and as a tangible actor in a conflict.

Now we reach the critique of the *consequences and unintended side-effects*: The abstraction from the individual's actual life relationships and sensitivities results in the determination of the individual as an abstract figure of thought and placeholder in varyingly constructed alternatives of a conflict in Mode C. Conflict models and problem solutions are thus transferred to the individual from the outside, i.e., from Mode C. This anonymized procedure de-individualizes, incapacitates, and defines the individual. As an individual in Mode A and Mode B, it should not be so easy to escape this disposition. It requires courage and impulses such as inversion. Because, on the one hand, according to G. Schulze, "['we'] behave as victims of circumstances because we are afraid to be active. The truth is: we are afraid of freedom" (Schulze 2006). On the other hand, we live in a time in which it is loudly expressed that "the discrediting of the 'dilettante' and the call for the specialist have taken the place of stubbornness and trial and error" (Schulze 2006). "The history of mankind began as a do-it- yourself project and led to dependence on experts" (Schulze 2006).

The individual in Mode A and Mode B, who has been brought to such a state of lack of independence and incapacity, is unlikely to be able to face the demands of a conflict on his own and to demand meaningful knowledge content from Mode C, according to the reference to *life-related pragmatic* criticism. Its critical faculty will atrophy and it will be absorbed in the 'given' (Adorno 1971; Kant 1784).

References

Adorno, T. W. (1971). *Erziehung zur Mündigkeit. Vorträge und Gespräche mit Hellmut Becker 1959 bis 1969*. Frankfurt (Main): Suhrkamp.

Allen, C. D. (2011). On Actor-Network Theory and landscape. *Area 43* (3), 274–280. https://doi.org/10.1111/j.1475-4762.2011.01026.x

Allen, J. (2008). Pragmatism and power, or the power to make a difference in a radically contingent world. *Geoforum 39* (4), 1613–1624. https://www.sciencedirect.com/science/article/pii/S0016718507000978.

Berger, P. L. & Luckmann, T. (1966). *The Social Construction of Reality. A Treatise in the Sociology of Knowledge*. New York: Anchor Books.

Bernstein, R. J. (2008). Richard Rorty's Deep Humanism. *New Literary History 39* (1), 13–27. https://muse.jhu.edu/article/240141.

Berr, K., Jenal, C., Koegst, L. & Kühne, O. (2022). *Noch mehr Sand im Getriebe? Kommunikations- und Interaktionsprozesse zwischen Landes- und Regionalplanung, Politik und Unternehmen der Gesteinsindustrie*. Wiesbaden: Springer VS.

Berr, K. & Kühne, O. (2023). Aesthetic constructions of landscape between society, individual and objects. A neopragmatic approach. *Studi di estetica LI, IV Serie* (26), 1–29. https://doi.org/10.7413/1825864638.

Berr, K. & Kühne, O. (2024). Landschaften sich schließender Gesellschaften. Eine Kritik am Leitfaden von Lebenschancen. In F. Bellin-Harder (Hrsg.), *Landschaft und Vegetation. Reflexionen zu Erkenntnistheorie, Herstellung und Interpretation* (Einzelveröffentlichungen Sozialwissenschaften, S. 147–180). Bielefeld: transcript.

Berr, K., Lohmann, P. & Kühne, O. (2023). The Contributions of Philosophy and the Social Sciences to Landscape Conflict Research—A Critical Comparison. *Sustainability 15* (24), 16802. https://doi.org/10.3390/su152416802

Brehm, J. W., Stires, L. K., Sensenig, J. & Shaban, J. (1966). The attractiveness of an eliminated choice alternative. *Journal of Experimental Social Psychology 2* (3), 301–313.

Davidson, D. (2004). *Subjektiv, intersubjektiv, objektiv*. Frankfurt (Main): Suhrkamp.

Dodge, M. & Perkins, C. (2008). Reclaiming the Map: British Geography and Ambivalent Cartographic Practice. *Environment and Planning A: Economy and Space 40* (6), 1271–1276. https://doi.org/10.1068/a4172

Dodge, M. & Perkins, C. (2015). Reflecting on J.B. Harley's Influence and What He Missed in "Deconstructing the Map". *Cartographica: The International Journal for Geographic Information and Geovisualization 50* (1), 37–40. https://doi.org/10.3138/carto.50.1.07

Edler, D. & Kühne, O. (2022). Deviant Cartographies: A Contribution to Post-critical Cartography. *KN – Journal of Cartography and Geographic Information 72,* 103–116. https://doi.org/10.1007/s42489-022-00110- w

Eisel, U. & Körner, S. (Hrsg.). (2009). *Befreite Landschaft. Moderne Landschaftsarchitektur ohne arkadischen Ballast?* (Beiträge zur Kulturgeschichte der Natur, Bd. 18). Freising: Technische Universität München.

Ellin, N. (1999). *Postmodern Urbanism*. New York: Princeton Architectural Press.

Flaßpöhler, S. (2021). *Sensibel. Über moderne Empfindlichkeit und die Grenzen des Zumutbaren*. Stuttgart: Klett-Cotta.

Fontaine, D. (2017). *Simulierte Landschaften in der Postmoderne. Reflexionen und Befunde zu Disneyland, Wolfersheim und GTA V*. Wiesbaden: Springer VS.

Gethmann, C. F. (2023). *Konstruktive Ethik. Einführung und Grundlegung*. Berlin, Heidelberg: Springer.

Grau, A. (2017). *Hypermoral. Die neue Lust an der Empörung* (2. Aufl.). München: Claudius.

Hagendorff, T. (2011). Ästhetische Wahrheitstheorien nach Rorty. *Critica – Zeitschrift für Philosophie und Kunsttheori 2,* 2–24.

Hilbig, H. (2024). Landschaft aus Akteur-Netzwerk-theoretischer Perspektive. In O. Kühne, F. Weber, K. Berr & C. Jenal (Hrsg.), *Handbuch Landschaft* (2. Aufl., 227-238). Wiesbaden: Springer VS.

Hubig, C. & Luckner, A. (2013). Klugheitsethik/Provisorische Moral. In A. Grunwald (Hrsg.), *Handbuch Technikethik* (S. 148–153). Stuttgart: J.B. Metzler.

Huizinga, J. (2019 [1938]. *Homo Ludens. Vom Ursprung der Kultur im Spiel* (26. Aufl.). Reinbek: Rowohlt. (Originalarbeit erschienen 1938).

Jackson, J. B. (1984). *Discovering the Vernacular Landscape*. New Haven: Yale University Press.

Jahncke, M. & Sprißler, N. (2024). An Actor-Network Theory Analysis of the Socio-spatial Landscape of Tübingen. In F. Papadimitriou & O. Kühne (Hrsg.), *Deviant Landscapes. A Journey to Exotic and Futuristic Landscapes and Spaces* (in this anthology). Wiesbaden: Springer VS.
Kant, I. (1784). Beantwortung der Frage: Was ist Aufklärung? *Berlinische Monatsschrift* (12), 481–494.
Kneer, G. & Nassehi, A. (1997). *Niklas Luhmanns Theorie sozialer Systeme. Eine Einführung.* München: Fink.
Koegst, L. (2022). Über drei Welten, Räume und Landschaften. Digital geführte Exkursionen an Hochschulen aus der Perspektive der drei Welten Theorie im Allgemeinen und der Theorie der drei Landschaften im Speziellen. *Berichte Geographie und Landeskunde 69* (3), 1–21. https://doi.org/10.25162/bgl-2022-0012
Korf, B. (2022). *Schwierigkeiten mit der kritischen Geographie. Studien zu einer reflexiven Theorie der Gesellschaft* (Sozial- und Kulturgeographie, Band 57). Bielefeld: transcript.
Kostner, S. (2019). Einleitung: Identitätslinke Läuterungsagenda. Genese des Analysekonzepts und Ziele des Bandes. In S. Kostner (Hrsg.), *Identitätslinke Läuterungsagenda. Eine Debatte zu ihren Folgen für Migrationsgesellschaften* (S. 7–17). Stuttgart: Ibidem-Verlag.
Kretz, D. (2023). Contingency, Irony, and Solidarity (1989). Übers.: Kontingenz, Ironie und Solidarität (1989). In M. Müller (Hrsg.), *Handbuch Richard Rorty* (1st ed. 2023, S. 201–224). Wiesbaden: Springer Fachmedien.
Kühne, O. (2006). *Landschaft in der Postmoderne. Das Beispiel des Saarlandes.* Wiesbaden: DUV.
Kühne, O. (2016). Transformation, Hybridisierung, Streben nach Eindeutigkeit und Urbanizing former Suburbs (URFSURBS): Entwicklungen postmoderner Stadtlandhybride in Südkalifornien und in Altindustrieräumen Mitteleuropas – Beobachtungen aus der Perspektive sozialkonstruktivistischer Landschaftsforschung. In S. Hofmeister & O. Kühne (Hrsg.), *StadtLandschaften. Die neue Hybridität von Stadt und Land* (S. 13–36). Wiesbaden: Springer VS.
Kühne, O. (2019). Sich abzeichnende theoretische Perspektiven für die Landschaftsforschung: Neopragmatismus, Akteur-Netzwerk-Theorie und Assemblage-Theorie. In O. Kühne, F. Weber, K. Berr & C. Jenal (Hrsg.), *Handbuch Landschaft* (S. 153–162). Wiesbaden: Springer VS.
Kühne, O. (2020). Landscape Conflicts. A Theoretical Approach Based on the Three Worlds Theory of Karl Popper and the Conflict Theory of Ralf Dahrendorf, Illustrated by the Example of the Energy System Transformation in Germany. *Sustainability: Science, Practice and Policy 12* (17), 1–20. https://doi.org/10.3390/su12176772
Kühne, O., Berr, K. & Jenal, C. (2023). *The Closed Society and Its Ligatures - A Critique Using the Example of 'Landscape'.* Wiesbaden: Springer Fachmedien; Springer.
Kühne, O., Berr, K. & Lohmann, P. (2023). *Landschaft zwischen Philosophie und Sozialwissenschaften. Eine Kritik.* Wiesbaden: Springer Fachmedien; Springer VS.
Kühne, O., Berr, K., Schuster, K. & Jenal, C. (2021). *Freiheit und Landschaft. Auf der Suche nach Lebenschancen mit Ralf Dahrendorf.* Wiesbaden: Springer.
Kühne, O., Jenal, C. & Edler, D. (2020). Functions of Landscape in Games – A Theoretical Approach with Case Examples. *Arts 9* (4). https://doi.org/10.3390/arts9040123
Kühne, O., Jenal, C. & Koegst, L. (2020). Postmoderne Siedlungsentwicklungen in Baton Rouge, Louisiana: Stadtlandhybridität und Raumpastiches zwischen Begrenzungen und Entgrenzungen. In F. Weber, C. Wille, B. Caesar & J. Hollstegge (Hrsg.), *Geographien der Grenzen. Räume – Ordnungen – Verflechtungen* (391–411). Wiesbaden: Springer VS.
Kühne, O. & Koegst, L. (2023). *Land Loss in Louisiana. A Neopragmatic Redescription.* Wiesbaden: Springer VS.
Kühne, O., Koegst, L. & Berr, K. (2024). *Oilscapes of Louisiana – Neopragmatic Reflections on the Ambivalent Aesthetics of Landscape Constructions.* Wiesbaden: Springer Fachmedien.
Longan, M. W. (2008). Playing with Landscape. Social Process and Spatial Form in Video Games. *Aether: The Journal of Media Geography* (2), 23–40.
Luhmann, N. (1984). *Soziale Systeme. Grundriß einer allgemeinen Theorie.* Frankfurt (Main): Suhrkamp.

Lyotard, J.-F. (1979). *La condition postmoderne. Rapport sur le savoir*. Paris: Les Éditions de Minuit.
Mittelstraß, J. (1974). *Die Möglichkeit von Wissenschaft*. Frankfurt (Main): Suhrkamp.
Mittelstraß, J. (1994). Der unheimliche Ort der Geisteswissenschaften. *Zweites Stuttgarter Bildungsforum. Orientierungswissen versus Verfügungswissen: Die Rolle der Geisteswissenschaften in einer technologisch orientierten Gesellschaft. Reden bei der Veranstaltung der Universität Stuttgart am 27. Juni 1994,* 30–39. herausgegeben von Ulrich Engler.
Mittelstraß, J. (1995). Der unheimliche Ort der Geisteswissenschaften. In U. Engler (Hrsg.), *Zweites Stuttgarter Bildungsforum. Orientierungswissen versus Verfügungswissen: Die Rolle der Geisteswissenschaften in einer technologisch orientierten Gesellschaft. Reden bei der Veranstaltung der Universität Stuttgart am 27. Juni 1994* (S. 30–39). Stuttgart: Universitätsbibliothek Stuttgart.
Nowotny, H. (2000). Sozial robustes Wissen und nachhaltige Entwicklung. *GAIA – Ecological Perspectives for Science and Society 9* (1), 93–100. https://doi.org/10.14512/gaia.9.1.1
Popper, K. R. (1996). *Alles Leben ist Problemlösen. Über Erkenntnis, Geschichte und Politik*. München: Piper.
Prominski, M. (2004). *Landschaft entwerfen. Zur Theorie aktueller Landschaftsarchitektur*. Berlin: Reimer.
Putnam, H. (1997). *Für eine Erneuerung der Philosophie* (Universal-Bibliothek, Bd. 9660). Stuttgart: Reclam.
Quast, C. (2021). *Was sind Experten? Eine begriffiche Grundlegung*. Frankfurt (Main): Campus.
Reckwitz, A. (2017). *Die Gesellschaft der Singularitäten. Zum Strukturwandel der Moderne*. Berlin: Suhrkamp.
Reckwitz, A. (2019). *Das Ende der Illusionen. Politik, Ökonomie und Kultur in der Spätmoderne*. Berlin: Suhrkamp.
Rogers, C. R. (2005). *Die klientenzentrierte Gesprächspsychotherapie. Client-Centered Therapy* (17. Aufl.). Frankfurt: Fischer.
Rorty, R. (1980). *Philosophy and the mirror of nature* (1. paperback print). Princeton, N. J.: Princeton University Press.
Rorty, R. (1997). *Achieving our country. Leftist thought in twentieth-century America* (The William E. Massey, Sr. lectures in the history of American civilization, Bd. 1997). Cambridge, Mass.: Harvard University Press.
(1997 [1989]. *Contingency, Irony, and Solidarity*. Cambridge: Cambridge University Press.
Rorty, R. (1998). The Contingency of Language. In M. F. Bernard-Donals & R. R. Glejzer (Hrsg.), *Rhetoric in an antifoundational World. Language, culture, and pedagogy* (S. 65–85). New Haven, Connecticut: Yale University Press.
Rorty, R. (1999). *Philosophy and social hope* (Penguin books, 1. publ). London: Penguin Books.
Rorty, R. (2023). *Pragmatismus als Antiautoritarismus*. Berlin: Suhrkamp.
Salaverría, H. (2023). Prophetische Zweifel und der „dunkel erahnte Zusammenhang von Kunst und Folter" – zur politischen Ästhetik Rortys. In M. Müller (Hrsg.), *Handbuch Richard Rorty* (1st ed. 2023, S. 933–947). Wiesbaden: Springer Fachmedien.
Schulz von Thun, F. (2019). *Miteinander reden. 1–4: Störungen und Klärungen/Stile, Werte und Persönlichkeitsentwicklung/Das „Innere Team" und situationsgerechte Kommunikation/Fragen und Antworten*. Hamburg: Rowohlt Taschenbuch.
Schulze, G. (Neue Züricher Zeitung, Hrsg.). (2006). Das süsse Gift der Entmündigung. https://www.nzz.ch/folio/das-susse-gift-der-entmundigung-ld.1619379. Zugegriffen: 28. Juni 2024.
Sloterdijk, P. (1987). *Kopernikanische Mobilmachung und ptolemäische Abrüstung. Ästhetischer Versuch* (Edition Suhrkamp). Frankfurt (Main): Suhrkamp.
Stintzing, M., Pietsch, S. & Wardenga, U. (2020). How to Teach "Landscape" Through Games? In D. Edler, C. Jenal & O. Kühne (Hrsg.), *Modern Approaches to the Visualization of Landscapes* (S. 333– 349). Wiesbaden: Springer VS.
Winsky, N. (2024). Landschaft in der Assemblage-Theorie. In O. Kühne, F. Weber, K. Berr & C. Jenal (Hrsg.), *Handbuch Landschaft* (2. Aufl., 217-226). Wiesbaden: Springer VS.

Chapter 7
Conclusion

According to the neopragmatist idea, the world is created and developed linguistically, and in this respect dealing with language is a prerequisite for the development of society. In our book, we have principally dealt with visual language, especially that of cartography. In doing so, we have made the effort to take up the criticism of critical cartography concerning traditional cartography, especially with regard to its power-bound nature and its positivist world view; instead of excluding (carto)graphic representations from the vocabulary, to promote their further development into a contingency-sensitive cartography. To this end, we have subjected the approach of representing inverse landscapes to critical reflection and further development. This approach is based on the idea of neopragmatism that the world is contingent and that increasing contingency leads to the development of new, useful solutions to problems. Accordingly, our aim was to maneuver between the Scylla of the diagnosis of all-embracing power-boundness with the consequence of total deconstruction (which implies the danger of social sclerosis), on the one hand, and the Charybdis of falling back into representational patterns of world interpretation on the other, which in extreme cases are based on the idea that ‘the world’ can be ‘objectively’ depicted (Rorty 1980). The emphasis on contingency does not eliminate the power-bound nature, in this case of representations, but power loses its potential to define interpretations as clearly legitimate. In particular, a (self-)ironic approach, as we have thematized here, has the potential to question power—but it also has the potential to exert power, for example over those who do not have the cultural capital to use the code of irony (here in particular in relation to cartography). At this point, the secondary consequences of human action become clear. However, the fact that there are both intended and unintended consequences as side-effects should not deter us from striving for innovation, but rather encourage us to anticipate and weigh up these consequences and side-effects and then to consider how dysfunctional effects can be minimized. This book is a document of such an approach. The aim was not only to further develop the concept of inverse landscapes, but also to expand its functional

O. Kühne et al., *Neopragmatism - Inverse Landscape - (Carto)graphic Representation*, Neopragmatic Horizontal Geographies 2, https://doi.org/10.1007/978-3-031-89402-2_7

and meta-functional potential and contain its dysfunctional potential as a result of critical considerations.

Our book draws on an expanded concept of life chances by Ralf Dahrendorf. If the goal of social development is to expand life opportunities, this is associated with an expansion of capabilities. This expansion can be outlined by dealing with two dimensions, one being the dimension of options, where the aim is not only to create individual options, but also individually available options. In the context of contingent and inverse landscape (and its representation): The option of being able to generate alternative designs of landscape, if necessary to be able to implement them in Space 1 interpreted as landscape (such as the area outside of one's own apartment/house). Regarding the second dimension, that of ligatures, the importance of ethical ligatures became clear; they are a prerequisite for examining options, yet contrastingly for voluntarily entering into ligatures. Moral ligatures, notably, especially if they are resentment-driven and externally oriented, have an effect of reducing contingency. In this respect, moral ligatures require ethical reflection. When dealing with contingent and inverse landscapes (and their representation), this means that ligatures must not serve to hinder the engagement with contingency in relation to space and landscape, but also the creation of contingent spatial constructions. Specifically, resentment-driven external moral ligatures have as an intended consequence or unintended side effect the loss of contingency. Instead of reducing contingency to achieve a concrete utopia or on the basis of the total deconstruction of the world, a neopragmatist approach to the world is characterized by enabling World 2 to understand contingency, in this case spatial and landscape contingency, as enrichment, not least to develop its own and self-responsible approach. This refers to the expression of a 'serene awareness of contingency': irony. The current challenge here is the unequal distribution of cultural capital, the prerequisite for developing an ironic attitude, yet also the ability to comprehend irony as an inauthentic world view on the other. Since neopragmatist irony should in no way serve as an instrument of social distinction, it is always incumbent on the 'liberal ironist' to restrict ironic communication to those social contexts in which irony is not understood as compromising. A further step—as already discussed above—can be seen in the dissemination of cultural capital, especially that which is related to orientational knowledge.

Inverse landscapes represent manifestations of contingent landscapes. With their representation in the form of images and cartographic representations, contingency is reduced, which initially appears to run counter to the concern of neopragmatism to broaden contingency. As a result of the widespread tendency of people to reduce contingency after a brief phase of interest, the representation of such inverse landscapes can be understood as an offer to accommodate 'quick reasoning' while still representing contingency. In other words: The concretization of contingency makes it possible to deal with the subject of contingency, even if the person already is actually on the way to creating unambiguity. To avoid misunderstanding inverse representations as 'concrete utopias', however, means of ironic cartography are also used—combined with the challenge discussed above of not appearing distinctive. In this respect, the pragmatic principle of orienting the depictions to the needs and skills of the target audience applies here again. The style of the ligne claire comics we use can also be linked to the activities of cartography (including traditional cartography): They represent an unconventional interpretation of the generalization customary in cartography. They reduce the number and variety of graphic objects—but not in an arbitrary way—and thus clarify the reference to cartography and concurrently to the contingency philosophy of neopragmatism.

In this book, we have deepened and differentiated the understandings and relationships of contingent landscapes and inverse landscapes on the basis of a critical exploration (more on this later):

1) We have made the manifestation of inverse landscapes from contingent landscapes more precise by specifying the relationships between landscape, not-landscape, and contingent landscape as well as introducing the Kippsaum/Pillepallerand.
2) We have made an instantiation of the contouring of contingent landscapes on not-landscapes against the background of different landscape construction patterns. In doing so, we have focused on the different ways in which Landscape 2 is constructed, i.e., that people can exhibit different levels of sensitivity to contingency, can have a high degree of contingency awareness in relation to landscape, yet can also reject contingency and strive for unambiguous classification. A wide transitional area can be found between the poles of contingency philia and contingency phobia.
3) We have developed a model of how the process of landscape-related communication can take place between ALTER and EGO by means of inverse landscapes. Thus, the area of what is referred to as 'landscape' can be widened by means of the notion of inverse landscapes, whereby the acceptance of the communication of inverse landscape can also be denied and
4) thus the understanding of landscape by EGO is not widened.
5) We have differentiated the contingent content of landscapes. This is based on the observation that the creation of landscape always relies on certain patterns that make a 'something' observable as landscape. This means that contingent as well as inverse landscapes rarely leave the realm of the 'landscape' in their entirety, as concerns the structures and processes that constitute them. Consequently, we assume that beyond the transition to the Kippsaum/Pillepallerand, primarily the development of landscapes that are partially contingent or partially inverse.
6) We have achieved a stronger integration of feeling and compassion in landscape research, in general, and in neopragmatist landscape research, in particular. The creation and communication of inverse landscapes and their experience is not limited to a cognitive confrontation, but also involves emotional affection. However, it is not only the creation of Landscape 2 that is partly based on feelings, but also the intersubjective communicative engagement with landscape is (partly) based on feelings. Empathic communication about landscape becomes possible when not only one's own sentiment-based relationship to landscape is communicated, but also when the creation of landscape by other people is subjected to appreciative attention by means of compassion.

7) We have also concretized and contoured the concept of neopragmatist landscape research. First, this concerns theoretical approaches that are completely incommensurable. Essentialist approaches are in no way compatible with neopragmatist thinking in their striving to find 'unambiguous truth' in the object. Attempts to make essentialist thinking fruitful for neopragmatism have led to the conclusion that the only fruitfulness that essentialist thinking can provide for neopragmatism lies in the clarification of the anti-essentialist position of neopragmatism. In addition, there are also positions that are partially incompatible with neopragmatism, such as utopianism, when it is not about the meta-utopia of the inclusive society. 'Critical theory' and deconstructivist approaches, which are less interested in solving concrete problems, while partially pursuing criticism as an end in itself, are also only partially compatible with neopragmatist thinking. A deconstructive approach can be used to understand specialized vocabularies and how they function. Neopragmatism, however, does not limit itself to determining the 'time-honored' state of vocabularies, but rather pursues their redescription by treating 'time-honored' vocabularies with respect. It also rejects the idea of a final vocabulary; its inclusive utopia does not refer to the design of vocabularies themselves, but to the relationships between vocabularies.
8) As a consequence of our engagement with the neopragmatist landscape theory, we also looked at the suitability of the propaedeutic basis we used, the theory of the three landscapes derived from Karl Popper's three world theory. Particularly against the background of communicative processes, here initially between ALTER and EGO, the need to expand the theory of the three landscapes to include the intersubjective became clear. With our 'horizontal' extension of the three world theory to include other worlds and their interaction with each other, we have created the prerequisite for being able to integrate the processes of intersubjective communication in negotiation processes around 'landscape' into our approach.

The starting point for these extensions, in the context of the neopragmatist understanding of landscape, was a criticism scheme developed elsewhere (Kühne et al. 2023). The critical approaches of internal criticism, which was able to point out internal inconsistencies, the criticism of consequences and unintended side effects—as it drew attention to possible dysfunctional effects of the approach—and the practical criticism, as these addressed the topic of the usefulness of neopragmatist landscape research in a special way, proved to be particularly fruitful for the further development of neopragmatist landscape theory in general, as well as the concepts of contingent and inverse landscapes. The ideological critique, having been already been mentioned, has led in particular to the question of what specifics a neopragmatist approach to landscape, how it differs from other approaches, to what extent a partial fruitfulness is possible, and which approaches are incommensurable (with the consequence that their critique can contribute little to the further development of our approach because they simply start from other basic assumptions, this especially applies to essentialist approaches). In dealing with the different types of critique, it also became clear that these should not be thought of as categorically distinct, but are characterized by hybridizations; for example, ideological critique can be thought of as critique from contextualization.

With our study, we have made a contribution to the operationalization of neopragmatist philosophy for spatial research by manifesting neopragmatist landscape theory in relation to the creation and communication of contingent landscapes and the production of inverse landscapes. These considerations are not only suitable towards a better understanding of the processes of the social and individual construction of landscape, but they are also suitable for being reflected in the practical handling of landscape. This practical significance is notable aimed at dealing with landscape change on all three levels. This becomes particularly significant against the background of changes to that space interpreted as landscape in the context of climate change. First, this relates to changes in ecosystems as a result of anthropogenic climate change (such as the immigration of autochthonous animal and plant species); second, this also relates to landscape-relevant adaptation measures to the changing climate (from the cultivation of other crops to the construction of technical flood protection measures); third, this also relates to the material manifestation of efforts to reduce greenhouse gas emissions (for example, through the construction of photovoltaic and wind power plants). Here, especially, the concept of inverse landscapes can contribute to acceptance research, for example, when it comes to understanding what changes are accepted, in which context, and what is not accepted. The creation of inverse landscapes manifest contingency. In this way, it can help to allay fears of contingency. At this point—as we have clearly worked out in the critique of the concept—there are the dangers of a relatively open utopianism aimed at reducing contingency (i.e., declaring a certain inverse landscape as binding), but also the danger of attempting to manipulate EGO through ALTER by choosing representations that (somewhat consciously) aim to create a desired impression. This cannot be ruled out, but by using the tool of ironic cartography, the attempt at manipulation can be counteracted.

As has become clear in our explanations, neopragmatism is characterized by a fundamental conceptual dilemma that lies between complex linguistic philosophy contrasted with the goal of finding solutions to practical problems. This also manifests itself in relation to dealing with contingent landscapes. Here, the claim (based on the philosophy of language) to increase contingency clashes with the everyday endeavor to outline the contingent as interesting, but to quickly arrive at an ordering unambiguity. At this point, inverse landscapes and their representations offer themselves as tools of translation; they can be interpreted as an intersection between contingency and efforts to actualize. Hence, they form a hinge (in relation to landscape) between the claims of the linguistic philosophy aspects of neopragmatism and the pragmatist portions. The (carto)graphic representations of inverse landscape are also indebted to this hinging function. They follow the principle of representationalism and thus contradict the basic idea of a more-than-representational theory, such as neopragmatism. But here, too, this contradiction is resolved partially by the ironic cartography, and unambiguity becomes undermined. However, the relationships underlying the hinging function are not stable, but are subject to processes of social change, in particular. In this respect, it is incumbent on a neopragmatist approach to the world to always deal self-critically with the appropriateness of the approach (here especially against the background of the increase in contingency contrasted by the contribution to finding practical solutions). Thus, from the metatheoretical to the action-oriented practice of life, it is always necessary to question one's own approach, to readjust it, to weigh up theoretical approaches, as well as methods and researcher perspectives.

Even if a critical approach has succeeded in concretizing neopragmatist landscape theory in general, the concepts of contingent and inverse landscapes, and their representations, there is still a need for further research in this context:

1) To date, neopragmatism has been used in spatial research particularly as a metatheory and as a basic normative understanding of an inclusive society. A pan-relationalist theory of space, region, and landscape has yet to be developed. Here it could—in comparison to other more- than-representational theories—play out its pragmatist heritage by not only examining networks and nodes, structures, and processes, but also asking how these could be designed in a more inclusive way.
2) This results in the challenge of developing a pan-relationalist cartography. This seems challenging since cartography has markedly cultivated representationalist thinking. This leads us to a further desideratum:
3) To date, forms of representation based on the comic style of the *ligne claire* have been developed. This form of representation is suitable for illustrating the idea of inverse landscapes, but only partially exploits both its creative and technical potential. In this respect, it is important a) to develop further forms of representation, b) to develop representation techniques, for example in the field of augmented and virtual reality.
4) As in other areas of science and everyday life, the question of how to deal with artificial intelligence also arises when dealing with contingent and inverse landscapes, especially their creation. As has been shown, artificial intelligence can produce representations of inverse landscapes very quickly. The question remains as to whether, and to what extent, these possibilities for the diverse production of representations of inverse landscapes have an influence on contingent landscapes, as well as on the interpretation of landscape and not- landscape—both individually and supra-individually. The question also arises as to how AI- generated inverse landscapes can be used meaningfully in landscape research and the practical handling of 'landscape'.
5) The dangers of these approaches were also addressed in our discussions of contingent and inverse landscapes. This also represents a largely unresolved research question: How to deal with utopian and manipulative inverse representations. This question can be approached historically by addressing the questions of how representations (which we can now describe as inverse) are produced in a utopian or manipulative way, but also what effect they had. However, this question can also be understood prospectively if we look at utopian or manipulative productions in the present and future. Here, too, AI has a significance that can hardly be estimated to date.

6) This (point 5) gives rise to a need for further research: How can a procedurally appropriate framework be designed with which the construction of inverse landscapes contributes to the expansion of contingency and is not used, or misused, to restrict contingency?
7) Last, but not least, we can return to our propaedeutic theory of the three landscapes: In this book, we have expanded it to include an intersubjective dimension. But intersubjective does not mean social. In this respect, the conceptual question here is how the social, which can be located between World 2 and World 3, can be integrated into the theory of the three landscapes (and also the three world theory).

References

Kühne, O., Berr, K. & Lohmann, P. (2023). *Landschaft zwischen Philosophie und Sozialwissenschaften. Eine Kritik*. Wiesbaden: Springer Fachmedien; Springer VS.

Rorty, R. (1980). *Philosophy and the mirror of nature* (1. paperback print). Princeton, N. J.: Princeton University Press.

MIX
Papier aus verantwortungsvollen Quellen
Paper from responsible sources
FSC® C105338

If you have any concerns about our products,
you can contact us on
ProductSafety@springernature.com

In case Publisher is established outside the EU,
the EU authorized representative is:
Springer Nature Customer Service Center GmbH
Europaplatz 3, 69115 Heidelberg, Germany

Printed by Libri Plureos GmbH
in Hamburg, Germany